T0205386

# Cognitive Systems Monographs

## Volume 35

**Series editors**

Rüdiger Dillmann, University of Karlsruhe, Karlsruhe, Germany
e-mail: ruediger.dillmann@kit.edu

Yoshihiko Nakamura, Tokyo University, Tokyo, Japan
e-mail: nakamura@ynl.t.u-tokyo.ac.jp

Stefan Schaal, University of Southern California, Los Angeles, USA
e-mail: sschaal@usc.edu

David Vernon, University of Skövde, Skövde, Sweden
e-mail: david@vernon.eu

*About this Series*

The Cognitive Systems Monographs (COSMOS) publish new developments and advances in the fields of cognitive systems research, rapidly and informally but with a high quality. The intent is to bridge cognitive brain science and biology with engineering disciplines. It covers all the technical contents, applications, and multidisciplinary aspects of cognitive systems, such as Bionics, System Analysis, System Modelling, System Design, Human Motion, Understanding, Human Activity Understanding, Man-Machine Interaction, Smart and Cognitive Environments, Human and Computer Vision, Neuroinformatics, Humanoids, Biologically motivated systems and artefacts Autonomous Systems, Linguistics, Sports Engineering, Computational Intelligence, Biosignal Processing, or Cognitive Materials as well as the methodologies behind them. Within the scope of the series are monographs, lecture notes, selected contributions from specialized conferences and workshops.

More information about this series at http://www.springer.com/series/8354

Mark Hoogendoorn · Burkhardt Funk

# Machine Learning
# for the Quantified Self

## On the Art of Learning from Sensory Data

 Springer

Mark Hoogendoorn
Department of Computer Science
Vrije Universiteit Amsterdam
Amsterdam
The Netherlands

Burkhardt Funk
Institut für Wirtschaftsinformatik
Leuphana Universität Lüneburg
Lüneburg, Niedersachsen
Germany

ISSN 1867-4925 ISSN 1867-4933 (electronic)
Cognitive Systems Monographs
ISBN 978-3-319-88215-4 ISBN 978-3-319-66308-1 (eBook)
https://doi.org/10.1007/978-3-319-66308-1

Printed on acid-free paper

This Springer imprint is published by Springer Nature
The registered company is Springer International Publishing AG
The registered company address is: Gewerbestrasse 11, 6330 Cham, Switzerland

*Live as if you were to die tomorrow.*
*Learn as if you were to live forever.*
Mahatma Gandhi

# Foreword

Sensors are all around us, and increasingly on us. We carry smartphones and watches, which have the potential to gather enormous quantities of data. These data are often noisy, interrupted, and increasingly high dimensional. A challenge in data science is how to put this veritable fire hose of noisy data to use and extract useful summaries and predictions.

In this timely monograph, Mark Hoogendoorn and Burkhardt Funk face up to the challenge. Their choice of material shows good mastery of the various subfields of machine learning, which they bring to bear on these data. They cover a wide array of techniques for supervised and unsupervised learning, both for cross-sectional and time series data. Ending each chapter with a useful set of thinking and computing problems adds a helpful touch. I am sure this book will be welcomed by a broad audience, and I hope it is a big success.

June 2017
Trevor Hastie
Stanford University, Stanford, CA, USA

# Preface

Self-tracking has become part of a modern lifestyle; wearables and smartphones support self-tracking in an easy fashion and change our behavior such as in the health sphere. The amount of data generated by these devices is so overwhelming that it is difficult to get useful insight from it. Luckily, in the domain of artificial intelligence, techniques exist that can help out here: machine learning approaches are well suited to assist and enable one to analyze this type of data. While there are ample books that explain machine learning techniques, self-tracking data comes with its own difficulties that require dedicated techniques such as learning over time and across users. In this book, we will explain the complete loop to effectively use self-tracking data for machine learning; from cleaning the data, the identification of features, finding clusters in the data, algorithms to create predictions of values for the present and future, to learning how to provide feedback to users based on their tracking data. All concepts we explain are drawn from state-of-the-art scientific literature. To illustrate all approaches, we use a case study of a rich self-tracking dataset obtained from the crowdsignals platform. While the book is focused on the self-tracking data, the techniques explained are more widely applicable to sensory data in general, making it useful for a wider audience.

Who should read this book? The book is intended for students, scholars, and practitioners with an interest in analyzing sensory data and user-generated content to build their own algorithms and applications. We will explain the basics of the suitable algorithms, and the underlying mathematics will be explained as far as it is beneficial for the application of the methods. The focus of the book is on the application side. We provide implementation in both Python and R of nearly all algorithms we explain throughout the book and make the code available for all the case studies we present in the book as well.

Additional material is available on the website of the book (ml4qs.org):

- Code examples are available in Python and R
- Datasets used in the book and additional sources to be explored by readers
- Up-to-date list of scientific papers and text books related to the book's theme

We have been researchers in this field for over ten years and would like to thank everybody who formed the body of knowledge that has become the basis for this book. First of all, we would like to thank the people at crowdsignals.io for providing us with the dataset that is used throughout the book, Evan Welbourne in particular. Furthermore, we want to thank the colleagues who contributed to the book: Dennis Becker, Ward van Breda, Vincent Bremer, Gusz Eiben, Eoin Grau, Evert Haasdijk, Ali el Hassouni, Floris den Hengst, and Bart Kamphorst. We also want to thank all the graduate students that participated in the Machine Learning for the Quantified Self course at the Vrije Universiteit Amsterdam in June 2017 and provided feedback on a preliminary version of the book that was used as reader during the course. Mark would like to thank (in the order of appearance in his academic career) Maria Gini, Catholijn Jonker, Jan Treur, Gusz Eiben, and Peter Szolovits for being such great sources of inspiration.

And of course, the writing of this book would not have been possible without our loving family and friends. Mark would specifically like to thank his parents for their continuous support and his friends for helping him in getting the proper relaxation in the busy book-writing period. Burkhardt is very grateful to his family, especially his wife Karen Funk and his two daughters, for allowing him to often work late and to spend almost half a year at the University of Virginia and Stanford University during his sabbatical.

Amsterdam, The Netherlands                                           Mark Hoogendoorn
Lüneburg, Germany                                                      Burkhardt Funk
August 2017

# Contents

# Chapter 1
# Introduction

Before diving into the terminology and defining the core concepts used throughout this book, let us first start with two fictive, yet illustrative, examples that we will return to regularly throughout this book.

The first example involves a person called *Arnold*. Arnold is 25 years old, loves to run and cycle, and is a regular visitor of the gym. His ultimate goal is to participate in an IRONMAN triathlon race consisting of 3.86 kilometers of swimming, 180 kilometers of cycling and running a marathon to wrap it all up—a daunting task. Besides being a fan of sports, Arnold is also a gadget freak. This combination of two passions has resulted in what one could call an obsession to measure everything around his physical state. He always wears a smart watch to monitor his heart rate and activity level and carries his mobile phone during all of his activities, allowing for his position and movements to be logged continuously in addition to a number of other measurements. He also installed multiple training programs on his mobile phone to help him schedule workouts. On top of that he uses an electronic scale in his bathroom that logs his weight and a chest strap to measure his respiration during running and cycling. All of this data provides him with information about his current state which Arnold hopes can help him to reach his ultimate goal making it to the finish line during the Hawaiian IRONMAN championship.

Contrary to Arnold, whom you could call a measurement enthusiast, *Bruce* also measures a lot of things around his body, but for Bruce this out of necessity. Bruce is 45 years old and a diabetic. In addition, he regularly falls into a depression. Bruce previously had trouble regulating his blood glucose levels using the insulin injections he has to take along with each meal. Luckily for Bruce, new measurement devices support him in to tackle his problems. He has access to a fully connected blood glucose measurement device that provides him with advice on the insulin dose to inject. To work on his mental breakdowns, Bruce installed an app that regularly asks him to rate his mental state (e.g. how Bruce is feeling, what his mood is, how well he slept, etcetera). In addition, the app logs all of his activities supported by location tracking and activity logging on his mobile phone, as it is known that a lack of activity

© Springer International Publishing AG 2018
M. Hoogendoorn and B. Funk, *Machine Learning for the Quantified Self*,
Cognitive Systems Monographs 35, https://doi.org/10.1007/978-3-319-66308-1_1

can lead to severe mental health problems. The app allows Bruce to pick up early signals on a pending mood swing and to make changes to avoid relapsing into a depression.

While Arnold and Bruce might be two rather extreme examples, they do illustrate the developments within the area of measurement devices: more and more devices are becoming available that measure an increasing part of our daily lives and well-being. Performing such measurements around one's self, quantifying one's current state, is referred to as the *quantified self*, which we will define more formally in the next section. This book aims to show how machine learning, also defined more precisely in this chapter, can be applied in a quantified self setting.

## 1.1  The Quantified Self

The term *quantified self* does not originate from academia, but was (to the best of our knowledge) coined by Gary Wolf and Kevin Kelly in Wired Magazine in 2007. Melanie Swan [114] defines it as follows:

**Definition 1.1** The quantified self is any individual engaged in the self-tracking of any kind of biological, physical, behavioral, or environmental information. There is a proactive stance toward obtaining information and acting on it.

When considering our two example individuals, Arnold would certainly be a quantified self. Bruce however, is not necessarily driven by a desire to obtain information, more by a better way of managing his diseases. Throughout this book we are not interested in this proactive stance, but in people that perform self-tracking with a certain goal in mind. We therefore deviate slightly from the definition provided before:

**Definition 1.2** The quantified self is any individual engaged in the self-tracking of any kind of biological, physical, behavioral, or environmental information. The self-tracking is driven by a certain goal of the individual with a desire to act upon the collected information.

What data precisely falls under the label quantified self is highly dependent on the rapid development of novel measurement devices. An overview provided by Augemberg [9] demonstrates the wealth of possibilities (Table 1.1). To what extent people track themselves varies massively, from monitoring the personal weight once a week to extremes that are inspired by projects such as the DARPA's LifeLog. For example, in 2004 Alberto Frigo started to take photos of everything he has used with his right hand, captured his dreams, songs he listened to, or people who he has met—the website 2004–2040.com is the mind-boggling representation of this effort.

Let us focus a bit on how widespread the quantified self is in society. Fox and Duggan [47] report that two thirds of US citizens keep track of at least one health indicator. Thus, following our definition, a large fraction of the US adult population

**Table 1.1**  Examples of quantified self data (cf. Augemberg [9], taken from Swan [114])

| Type of measurement | Examples |
| --- | --- |
| Physical activities | miles, steps, calories, repetitions, sets, METs (metabolic equivalents) |
| Diet | calories consumed, carbs, fat, protein, specific ingredients, glycemic index, satiety, portions, supplement doses, tastiness, cost, location |
| Psychological states and traits | mood, happiness, irritation, emotions, anxiety, self-esteem, depression, confidence |
| Mental and cognitive states and traits | IQ, alertness, focus, selective/sustained/divided attention, reaction, memory, verbal fluency, patience, creativity, reasoning, psychomotor vigilance |
| Environmental variables | location, architecture, weather, noise, pollution, clutter, light, season |
| Situational variables | context, situation, gratification of situation, time of day, day of week |
| Social variables | influence, trust, charisma, karma, current role/status in the group or social network |

belongs to the group of quantified selves. Even if we restrict our definition to those who use online or mobile applications or wearables for self tracking, the number of users is high: An international consumer survey by GfK [50] in 16 countries states that 33% of the participants (older than 15 years) monitor their health by electronic means, China being in the lead with 45%. There are many indicators that the group of quantified selves will continue to grow, one is, the number of wearables that is expected to increase from 325 million in 2016 to more than 800 million in 2020 [110].

What drives these quantified selves to gather all this information? Choe et al. [38] interviewed 52 enthusiastic quantified selves and identified three broad categories of purposes, namely to improve health (e.g. cure or manage a condition, achieve a goal, execute a treatment plan), to enhance other aspects of life (maximize work performance, be mindful), and to find new life experiences (e.g. learn to increasingly enjoy activities, learn new things). A similar type of survey is presented in [51] and considers self-healing (help yourself to become healthy), self-discipline (like the rewarding aspects of the quantified self), self-design (control and optimize yourself using the data), self-association (enjoying being part of a community and to relate yourself to the community), and self-entertainment (enjoying the entertainment value of the self-tracking) as important motivational factors for quantified selves. They refer to these factors as "Five-Factor-Framework of Self-Tracking Motivations".

While Gimple et al. [51] study the goals behind the quantified self, Lupton [83] focus on what she calls modes of self-tracking and distinguishes between private and pushed self-tracking, the latter referring to situations in which the incentive to engage in self-tracking does not come from the user himself but another party. This being said, not only users themselves are interested in the data generated within the context of the quantified self. Health and life insurances come to one's mind immediately, they love to know as much as possible about the current health status and lifestyle of a potential customer before underwriting an insurance contract. For insurance companies, leveraging self-tracking data for personalized offerings is a natural next step to questionnaire based assessments that currently employed. Insurers do not have to force their customers to share their data, but can set financial incentives to do so. Besides insurances and health providers, other companies are also keen to tap into this data source. Companies, e.g. from the recreation industry, like to understand user behavior and location to target their offerings. Only recently, "the workplace has become a key site of pushed self-tracking, where financial incentives or the importance of contributing to team spirit and productivity may be offered for participating" [83].

Since self-tracking data can be misused or used in a way that is not fully in the interest of a person, it is not surprising that users state the loss of privacy as their main concern in this context. For example, in 2013 it was reported that a supermarket chain in the UK used wearables to monitor their employees who in return (and again not surprising) felt a lot of pressure. As said before, user profiling with respect to health and fitness behavior will help companies to personalize their offerings. For some users this might be beneficial, others might be excluded as customers, as is obvious in the insurance and financial industry. Another very sensitive piece of the quantified self data is location that can be abused for criminal purposes but also to increase control by public authorities.

We are aware that an intensive, open, and broad discourse on self-tracking is needed that puts the interest of individuals first. However, to discuss these risks, personal concerns, and also the opportunities that come with the quantified self for individuals and companies is far beyond the more technical and methodological perspective of our book. A good starting point for this discussion is the book by Neff and Nafus [89].

## 1.2   The Goal of this Book

Now that we know more about the quantified self, what do we seek to achieve with this book? As you might have noticed, the quantified self can and will most likely result in a huge amount of data being collected about individuals. An immediate question that pops up is how to make sense of this data. Even enthusiasts such as Arnold will not be able to oversee it all, and might miss valuable information. This is where machine learning comes into play. Many definitions of machine learning exist. In our case, we define machine learning as follows:

**Definition 1.3** Machine learning is to automatically identify patterns from data.

This book aims at showing how machine learning can be applied to quantified self data; specifically to automatically extract patterns from collected data and to enable a user to act upon insights effectively, which in turn contributes to the goal of the user. Let us make this a bit more concrete for our two fellows Arnold and Bruce by illustrating potential situations and questions:

- Advising the training to make most progress towards a certain goal based on past training outcomes
- Forecasting when a specific running distance will be feasible based on the progress made so far and the training schedule
- Predict the next blood glucose level based on past measurements and activity levels
- Determine when and how to intervene when the mood is going down to avoid a spell of depression
- Finding clusters of locations that appear to elevate one's mood

All these questions could be answered by extracting patterns from historical data.

An observant reader might ask at this point whether this is yet another book in the area of machine learning among many others. The data from the quantified self does however pose its own challenges, which requires dedicated algorithms and data preparation steps. We will precisely focus on this area and take a more applied stance. For more theoretical underpinning of algorithms the reader will be referred to fundamental machine learning books such as Hastie et al. [57] and Bishop [18].

So what are the unique characteristics of machine learning in the quantified self context? We identify five of them: (1) sensory data is noisy, (2) many measurements are missing, (3) the data has a highly temporal nature, (4) algorithms should enable the support of and interaction with users without a long learning period, and (5) we collect multiple datasets (one per user) and can learn across them. Each of these issues will be treated in this book. Note that the approaches we introduce here are not limited to the development of applications for quantified selves, but that are also relevant for a broader category of applications, such as predictive modeling for electronic medical record data (think of a patient lying at the ICU for example).

## 1.3 Basic Terminology

Before explaining the formal notation used throughout this book, we will introduce some terminology first. This is by no means meant to be complete, but will provide a basic vocabulary that we can build upon. We will start with the introduction of basic terms to describe aspects of data, followed by some basic machine learning terminology.

### 1.3.1 Data Terminology

Datasets encompass different attributes such as the heart rate of a person or the number of steps per day. The most elementary part of data is in our case a measurement, which is defined as follows:

**Definition 1.4** A measurement is one value for an attribute recorded at a specific time point.

Measurements can have values of different data types; they can be *numerical*, or *categorical* with an ordering (*ordinal*) or without (*nominal*). Let us consider an example dataset associated with Arnold. The attributes are shown in Table 1.2. The time point is not considered to be part of the attributes (though listed for the sake of completeness) as it is an inherent part of the measurement itself. For the other variables, the speed and heart rate would be considered a numerical measurement. The Facebook posts and activity type are both nominal attributes and the activity level is ordinal.

Measurements frequently come in sequences, for instance a sequence of values for the heart rate. This is what we call a *time series*:

**Definition 1.5** A time series is a series of measurements in temporal order.

Time series often form the basis to interpret measurements. To exemplify the notion of a time series, an example of data collected for each of the attributes discussed

**Table 1.2** Attributes in example dataset

| Time point | The time point at which the measurement took place (considered in hours for this example) |
|---|---|
| Heart rate | Beats per minute, integer value |
| Activity level | Can be either low, medium or high |
| Speed | Speed in kilometers per hour, real value |
| Facebook post | A string representing the Facebook message posted |
| Activity type | The type of activity: inactive, walking, running, cycling, gym |

**Table 1.3** Example dataset

| Time point | Heart rate | Activity level | Speed | Facebook post | Activity type |
|---|---|---|---|---|---|
| 14:30 | 55 | low | 0 | getting ready to hit the gym | inactive |
| 14:45 | 55 | low | 0 | having trouble getting off the couch | inactive |
| 15:00 | 70 | medium | 5 | walking to the gym, it's gonna be a great workout, I feel it | walking |
| 15:10 | 130 | high | 0 | - | gym |
| 15:50 | 120 | high | 12 | the gym didn't do it for me, running home | running |
| 16:15 | 130 | high | 35 | still have energy, on my bike now | cycling |

in Table 1.2 is shown in Table 1.3. In the table, the columns represent the attributes while the rows are the measurements performed at the indicated time points. Here, one can consider the sequence [55, 55, 70, 130, 120, 130] as an example of a time series for the attribute heart rate.

Now that we know the basic data terminology, let us move to the terminology of machine learning.

## *1.3.2  Machine Learning Terminology*

The field of machine learning is commonly divided into four types of learning problems: *supervised learning, unsupervised learning, semi-supervised learning*, and *reinforcement learning*. Except for semi-supervised learning, all these types of learning will be explored throughout this book in the context of the quantified self. Let us look at them in a bit more detail. First, consider the definition of supervised learning we adopt:

**Definition 1.6** Supervised learning is the machine learning task of inferring a function from labeled training data (cf. [87]).

Let us return to the example of the dataset depicted in Table 1.3. An example of a supervised learning problem would be to learn a function that determines the activity type based on the other measurements at that same time point. Here, each row in the table is a training example where the *label* (also known as the *target* or *outcome*) is the activity type. We will refer to an individual training example as an *instance* to stay in line with standard machine learning terminology. Attributes are also referred to as *variables* or *features*. We will use these terms interchangeably. Different types of supervised learning exist, which mainly depend on the type of variable that is being predicted. *Classification* is the term used in case the predicted type of data is categorical (e.g. the activity type for our example dataset) while *regression* is used for numerical measurements (e.g. the heart rate).

Moving on to another type of learning problem, unsupervised learning is the opposite of supervised learning:

**Definition 1.7** In unsupervised learning, there is no target measure (or label), and the goal is to describe the associations and patterns among the attributes (cf. [57]).

Examples of tasks within unsupervised learning that are considered in this book are *clustering* and *outlier detection*. Since there is no desired outcome (or "teacher") available, these algorithms typically try to characterize the data, and make assumptions about certain properties of this characterization. For clustering, the algorithm tries to group *instances* that share certain common characteristics given a definition of similarity. For our example dataset, you might find a cluster of intense activities and one with limited activities. In outlier detection, it is the goal to find points that appear to deviate markedly from other members of the sample in which it occurs.

The third type of learning, semi-supervised learning [33], combines the supervised and unsupervised approach of learning:

**Definition 1.8** Semi-supervised learning is a technique to learn patterns in the form of a function based on labeled and unlabeled training examples.

Since generating labeled training examples can take significant efforts, semi-supervised learning also makes use of unlabeled training examples to learn a target function. For example, assume we want to infer the mood of a user based on his smartphone usage patterns. To come up with a set of labeled training examples you would need to require the user to manually record his mood for a few weeks, which obviously is associated with some effort. Without too much effort, you might at the same time collect data on smartphone usage for other time periods for which you do not have mood ratings, an unlabeled set that could still provide a valuable contribution to the learning task. In many cases (e.g. face, speech, or object recognition) we have only few labeled training examples and vast amounts of unlabeled training data (think of all photos available on the Internet). That is why semi-supervised learning is currently an important topic in machine learning.

Finally, we consider reinforcement learning. The definition we use is similar to [112]:

**Definition 1.9** Reinforcement learning tries to find optimal actions in a given situation so as to maximize a numerical reward that does not immediately come with the action but later in time.

In reinforcement learning, the learner is not told which actions to take as in supervised learning but instead must discover which actions yield the highest reward over time by trying them. We can see that this is a bit different from our previous categories as we no longer immediately know whether we are right or not (like supervised learning) but we do in the end get a reward signal which we want to optimize given a *policy* (which specifies when to do what). For Arnold, a reward could for instance be an improvement of his long-term shape while the action that we try to learn is to give appropriate daily advice depending on Arnold's state.

## 1.4  Basic Mathematical Notation

While we focus more on applying techniques rather than explaining all of the fundamentals, we do aim to provide understanding of the algorithms to a certain extent. To provide this understanding, a consistent mathematical notation can greatly assist us. This is introduced in this section. In our mathematical notation, we use the same notation as introduced by Hastie et al. [57]. As a basic starting point, the input variables are denoted by $X$. Here, $X$ could be (and most likely is) a vector containing multiple variables. We assume that there are $p$ such variables. Think of our previous example where we aimed to predict the activity type. The inputs were heart rate, activity level, speed, and the Facebook post text. Each of the individual $p$ variables can be accessed by a subscript, i.e. for the $k$th variable $X_k$. For instance, $X_1$ denotes the variable heart rate in our example. In the case of supervised learning, the outputs will be denoted by $Y$ for regression problems or $G$ for classification. When there are multiple variables to predict we will again use a subscript to identify one specific variable. An observation of $X$—that is, a single instance of the data (with the observed values for all variables)—is denoted in lowercase: $x_j$. It represents a column vector of observations of our $p$ variables where $j$ identifies the instance. $j$ can take the values $j = 1, \ldots, N$ with $N$ being the number of observations. For example:

$$x_1 = \begin{bmatrix} 0 \\ 45 \\ \text{low} \\ 0 \\ \text{"getting ready to hit the gym"} \end{bmatrix}$$

If we want to refer to a specific value of one variable within the instance we will use the notation $x_j^k$ where $j$ refers to the instance and $k = 1, \ldots, p$ ($p$ is the number of variables) to the position of the variable in the vector (e.g. $x_1^2 = 45$). Here, depending on the nature of the instances, $j$ could also represent the notion of time as the instances might form a sequence of measurements over time, i.e. $j = t_{start}, \ldots, t_{end}$ assuming a discrete time scale. Given that we have $p$ elements in our vector, we can represent an entire dataset as a matrix (similar to the table notation we have seen before). This will result in an $N \times p$ matrix. As $x_j$ is defined to be a column vector (our example $x_1$ was as well) each row $j$ is the transposed version of $x_j$, i.e. $x_j^T$. This matrix will be noted in boldface with $\mathbf{X}$. Sometimes we use an index to identify a specific dataset (e.g. the dataset originating from Arnold or Bruce), we note this as $\mathbf{X_i}$. If the instances represent a sequence of measurements over time we will use $\mathbf{X}^T$ to denote a time series training set (this will be an important distinction for later chapters). If we omit the $\mathcal{T}$ we make no assumption about the ordering. The same conventions as we have just introduced are used for the targets for the case of supervised learning. The entire set of targets for all instances are specified by $\mathbf{Y}$ and $\mathbf{G}$ for numerical and categorical targets respectively. We have very distinct cases for numerical and categorical cases as the learning algorithms for both cases typically work very differently. The predicted output of our supervised model over all instances will be denoted as $\hat{\mathbf{Y}}$ or $\hat{\mathbf{G}}$. Individual targets and predictions for the instance $i$ are expressed as $y_i$ and $g_i$ for the target values and $\hat{y}_i$ and $\hat{g}_i$ for the predictions. Our target output for our input vector $x_1$ would be:

$$g_1 = \begin{bmatrix} inactive \end{bmatrix}$$

Hence, we end up with a training dataset of the form $(x_j, y_j)$ or $(x_j, g_j)$ where $j = 1, \ldots, N$. An overview of the notation is presented in Table 1.4.

**Table 1.4**  Mathematical notation

| Notation | Explanation |
|---|---|
| *Dataset representation* | |
| $X_k$ | A variable (or attribute) in our dataset, $k$ is the index of the variable |
| $\mathbf{X}_i^T$ | Matrix representing a dataset containing $N_i$ instances with $p$ variables. The $i$ allows us to refer to a specific dataset (e.g. of a specific person) while the $T$ indicates a dataset with a temporal ordering. If $T$ is omitted no assumption about the ordering within the dataset is made |
| $x_j^k$ | The $j$th observation in the dataset. $k$ refers to the specific variable within the observation. If $k$ is omitted this concerns an observation of the entire vector of variables |
| *Categorical target representation (optional)* | |
| $G$ | A categorical target variable in our dataset |
| $\mathbf{G}$ | Similar to $\mathbf{X}_i^T$ (and the same additional super- and subscripts can be used), except that this refers to the categorical targets for our dataset (if present). It contains $N$ instances |
| $g_j$ | The $j$th instance of the categorical target or row in $\mathbf{G}$ |
| *Classifier prediction representation* | |
| $\hat{g}_j$ | The prediction of our classifier of the target for the $j$th row in the dataset |
| $\hat{\mathbf{G}}$ | The entire set of categorical predictions of our classifier |
| *Numerical target representation (optional)* | |
| $Y$ | A numerical target variable in our dataset |
| $\mathbf{Y}$ | Similar to $\mathbf{X}_i^T$ (and again the same additional super- and subscripts can be used), except that this refers to the numerical targets for our dataset (if present). It contains $N$ instances |
| $y_j$ | The $j$th instance of the numerical target or row in $\mathbf{Y}$ |
| *Numerical prediction representation* | |
| $\hat{y}_j$ | The prediction of our model of the numerical targets for the $j$th row in the dataset |
| $\hat{\mathbf{Y}}$ | The entire set of numerical predictions of our model |

## 1.5   Overview of the Book

Figure 1.1 shows the main aspects of the book. The yellow box encompasses applications that collect data about the quantified self in various ways: user responses to questionnaires posed to the user in a certain context (ecological momentary assessment), data on usage behavior, data from physical sensors (think of an accelerometer), and audiovisual information obtained through cameras or microphones. Additional sensors which are not part of a smartphone or a wearable can also provide data. Examples are indoor positioning sensors, weather forecasts, or the medical history of a person. To use all of this data we need to do some pre-processing before we can actually perform the machine learning tasks we aim to do. This is indicated by the red box. Smoothing of the data, handling missing values and outliers, and the generation of useful features are the core aspects in this context. Based on the resulting dataset,

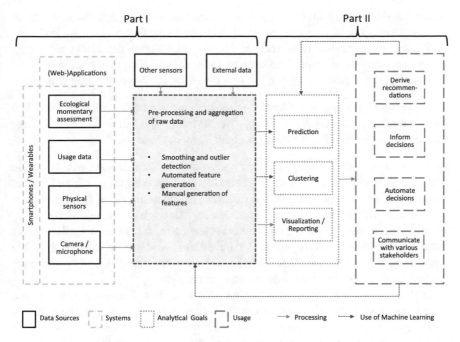

**Fig. 1.1** Various elements relevant to make sense out of quantified self data

we can perform varying types of analyses, e.g. : create models that can be used for prediction of unknown values using a variety of machine learning techniques, detect interesting patterns and relations in the data (e.g. clusters), and create visualizations to gain insights into the data. These analytical goals are shown in the green box. Finally, we can start using the knowledge we have gained (the blue box) in order to derive recommendations, inform decisions, and automate and communicating them with various stakeholders (in the context of Bruce, think of Bruce himself, his therapist, etc.). In accordance with this overview, this book has been divided into three main parts:

- The first part covers the pre-processing of the data and feature generation. We will start by explaining the basics of sensory data and introduce the dataset we use as a case study throughout nearly all chapters. Next, we explain how to smooth the data and remove obvious outliers. Finally, we will go into depth on the extraction of useful features from the cleaned data.
- The second part explains all relevant machine learning techniques that can help us to reach our analytical goals and also allow us to "close the loop", i.e. help us to use the outcomes of the analysis to support the user more effectively. The first topic we will cover is the clustering of the data. Here, we will focus on clustering of the data of a single user, but also the clustering on a higher level, namely the clustering over different users. We will then elaborate on the theoretical foundations behind

supervised learning, and cover supervised machine learning approaches, both those that exploit the temporal dimension of data and those that do not. We conclude with an introduction of reinforcement learning techniques that allow us to learn how to effectively intervene and support a user in achieving his or her goals.
- Finally, the third part is a discussion about avenues for future developments.

With this book, we aim for different target audiences, and we want to provide a reader's guide for the different groups. We have identified three target audiences (please do not be offended if you do not match any of these profiles):

- Scholars and students without prior background in machine learning: we would suggest to read the whole book to get up to speed on both the machine learning techniques and all specific issues concerning the quantified self. In case you are unfamiliar with the mathematical notations used throughout the book, we recommend [42], pages 601–635 as a brief overview of useful mathematical concepts before reading the book. If you want to have a light read and do not care about the principles behind the techniques we have shown, you can also just consider the introductions of the various chapters and the case study.
- Scholars and students with prior background in machine learning: for those who are familiar with the concepts within the field of machine learning, we certainly recommend part I, but would recommend the reader to skip the explanation of the basic machine learning techniques in Chap. 7, which will already be familiar to you. The learning setting (initial section of the chapters) and the case study are still very relevant.
- Professionals: if you are a professional who is more focused on the implementation of quantified self applications which embed machine learning techniques, we recommend reading the basis introductions of the different techniques and mainly focus on the case study and the associated code. The case study follows all our recommendations to develop a successful application.

Throughout the book, we make extensive use of a case study. All the code that we have written related to the case study is available on a per-chapter basis, both in Python and in R. It covers nearly all the algorithms we explain in the book. The code of the examples we use to explain the basics of the machine learning algorithms is available as well. All code can be found on the website.[1] We also provide exercises, which can be found at the end of every chapter throughout the book. These are questions about the chapter, but also about techniques we did not cover but consider to be relevant.

---

[1]ml4qs.org.

# Part I
# Sensory Data and Features

# Chapter 2
# Basics of Sensory Data

Before we discuss some of the details of the various machine learning approaches, we will focus on the topic of sensory data itself. Since rapid technical advances are being made in this area, we will refrain from explaining the workings of each potentially useful sensor out there. Rather, we will dive into a representative dataset used throughout the book. The dataset originates from *crowdsignals*[1] which has generously been made available for experimentation for us as authors and for you as reader of our book. The dataset has been collected using an application that gathers data from both a smartphone and a smart watch. In addition, users were asked to label the activities they were conducting (e.g. "I am currently running"). We will first describe the measurements included in the dataset. We will then show how to move from the raw data we collect to a dataset usable in machine learning tasks. This process is described in the context of the crowdsignals dataset but is representative for most of the sensory datasets we have worked with. Finally, we explore the resulting dataset and identify suitable machine learning tasks.

## 2.1 Crowdsignals Dataset

An overview of the sensory data in the crowdsignals dataset is shown in Table 2.1. In the table, we focus on the *sensors* and *user labels* categories, for the others, please explore the full crowdsignals dataset description, which is available via the aforementioned website. We were not able to include all sensor and user label measurements in the experiments we present in this book. Those that have been included are marked with a "yes" in the last column.

---

[1] http://www.crowdsignals.io.

© Springer International Publishing AG 2018

M. Hoogendoorn and B. Funk, *Machine Learning for the Quantified Self*,
Cognitive Systems Monographs 35, https://doi.org/10.1007/978-3-319-66308-1_2

**Table 2.1** Sensors and labels in crowdsignals dataset

| Sensor | Purpose | Device(s) | Values | Time point / Interval | Used |
|---|---|---|---|---|---|
| *Sensors* | | | | | |
| Accelerometer | The acceleration of the device | phone/ watch | x, y, and z acceleration | time point | yes |
| Gyroscope | The angular speed of the device | phone/ watch | x, y, and z angular speed | time point | yes |
| Magnetometer | The magnetometer value of the device | phone/ watch | x, y, and z magnetometer value | time point | yes |
| Heart rate | The heart rate of the user | watch | heart rate (beats per minute) | time point | yes |
| Temperature | Ambient temperature | phone/ watch | temperature (in °C) | time point | no |
| Light | The light intensity | phone/ watch | light intensity (in lux) | time point | yes |
| Pressure | The current pressure | phone/ watch | pressure (in mercury millibars) | time point | yes |
| Humidity | The current humidity | phone/ watch | relative humidity (%) | time point | no |
| Proximity | Distance of user from phone | phone | distance (meters) | time point | no |
| Audio record | Record of audio obtained via the microphone | phone | audio recording | time point | no |
| *User labels* | | | | | |
| Activity label | Record of the activity a user is conducting | phone | label (walking, running, ....) | interval | yes |

A huge variety of sensors exist. Three popular sensors do dominate the landscape of smartphone sensors and are also included in our dataset: the accelerometer, magnetometer, and gyroscope. The *accelerometer* measures the changes in forces upon the phone on the x, y, z-plane. The orientation of the phone compared to the "down" direction (the earth's surface) and the angular velocity are measured by means of the *gyroscope* (measured on the same three axes as the accelerometer does). Finally, the *magnetometer* measures the x-, y-, and z-orientation relative to the earth's magnetic field. Micro-electromechanical systems (MEMS) form the technical basis of these sensors. MEMS employ the effect that the resistance of semiconductors is stress-sensitive, or put in other words, changes when mechanical forces are applied—this phenomenon discovered in the 1950s is called piezoresistance and the basis of a large industry today [22].

**Table 2.2** Snapshot heart rate data

| Sensor_type | Device_type | Timestamps | Rate |
|---|---|---|---|
| heartrate | smartwatch | 1454956086325639687 | 175 |
| heartrate | smartwatch | 1454956086684549167 | 176 |
| heartrate | smartwatch | 1454956087523516770 | 175 |

**Table 2.3** Snapshot label data

| Sensor_type | Device_type | Label | Label_start | Label_end |
|---|---|---|---|---|
| interval_label | smartphone | On Table | 1454956132985999872 | 1454956366574000128 |
| interval_label | smartphone | On Table | 1454956393088000000 | 1454956578385999872 |
| interval_label | smartphone | On Table | 1454956608515000064 | 1454956813323000064 |
| interval_label | smartphone | Sitting | 1454956894057999872 | 1454957092968000000 |

Of course, there are many more sensors used in today's smartphones that you are familiar with. Just think of a GPS signal that measures your position by means of your distance to a number of satellites of which the position is known. For a full overview of sensors, we refer the reader to books dedicated to modern sensors, for example [48].

Let us have a look at how the data has been recorded. All data is stored with a reference to when the data was measured. Some recordings cover measurements for a certain period or interval while others are only valid for a specific point in time. For example, the heart rate is measured for a specific time point while the label provided by the user is specified for an interval (I was walking between time point $t$ and time point $t'$). In Table 2.2, we can see a snapshot of the heart rate data, whereas an example for the label data is shown in Table 2.3. Time points are expressed in nanoseconds since the start of time (which is January 1st 1970 following the UNIX convention).

We are still far away from the specification of a dataset we have seen in Chap. 1, where $\mathbf{X}^T$ denotes a matrix with rows representing the measurements of an individual time point (if the dataset has a temporal nature, which we clearly have here). Next, we will show how we move from our current dataset to the desired matrix format.

## 2.2 Converting the Raw Data to an Aggregated Data Format

In order to convert the temporal data, we first need to determine the time step size we are going to use in our dataset. This is also referred to as the level of granularity (selecting a $\Delta t$). We could say that we want to have instances covering a second of data for example, or even a minute. The selection of the step size depends on a variety of factors, including the task, the noise level, the available memory and cost of storage, the available computational resources for the machine learning process, etcetera. Once we have selected this step size we can create an empty dataset.

We start with the earliest time point observed in our crowdsignals measurements and generate a first row $x_{t_{start}}$. Iteratively, we create additional rows for the following time steps by taking the previous time step and adding our step size, e.g. $x_{t_{start}+\Delta t}$. Each row $x_t$ represents a summary of the values encountered in the interval defined by the time step it was created for until the next time step, i.e. $[t, t + \Delta t)$. We continue until we have reached the last time step in our dataset. Next, we should identify the columns in our dataset (our attributes) that we want to aggregate. As we have seen, we can distinguish between numerical values (e.g. the heart rate) and categorical values (e.g. the labels) and need different approaches for both. For the former, we create a single column for each variable we measure while for the categorical values we create a separate column for each possible value. Of course, for the categorical attributes we could also include a single column where each row would contain a single value for that measurement. However, since we are discretising time steps it is very likely that we will encounter multiple values for our categorical measurement per time step (e.g. the user performing the activity driving and walking within the same time step). We cannot accommodate for this if we can just insert a single value: which one should we select?

Once we have defined the entire empty dataset, we are ready to derive the values for each attribute at each discrete time step (i.e. each row). We select the measurements in our crowdsignals data that belong to the specific discrete time step (when either the associated time stamp falls in the window, or the interval expressed falls (partly) within it) and aggregate the relevant values. We can aggregate numerical values by *averaging* the relevant measurements (e.g. for heart rate) or we can *sum* them up (e.g. when the measurements concern a quantity) or use other descriptive metrics from statistics such as median or variance. Since often it is not clear a priori which type of aggregation to choose, you could also use different measures and later let machine learning techniques select relevant features. For categorical values we can count whether at least one measurements of that value has been found in the interval (*binary*) or we can count the number of measurements that have been found for the value (*sum*).

In our case we have selected the averaging method for numerical values and the binary method for categorical attributes. When taking a $\Delta t$ of 1 day and aggregating the data we have seen in Tables 2.2 and 2.3 we would end up with the table shown in Table 2.4. As mentioned before, all these approaches have been implemented and are available on the website accompanying the book, including the code used to process the crowdsignals dataset.

**Table 2.4**  Example resulting dataset

| Time | Heart_rate | Label On Table | Label Sitting |
|------|-----------|----------------|---------------|
| 2016-02-08 19:28:06 | 175.333 | 1 | 1 |
| 2016-02-09 19:28:06 | - | 0 | 0 |

## 2.3   Exploring the Dataset

Let us consider the entire dataset with the sensors we have marked as "yes" in Table 2.1. We have a set of measurements that covers approximately two hours of labeled data of a participant. If we take a granularity of 1 minute, we obtain a dataset that is shown in Fig. 2.1. The dataset contains 133 instances (i.e. 133 minutes). We can see that we have quite a nice dataset, although the data does seem a bit too smooth, especially regarding the accelerometer, gyroscope, and the magnetometer

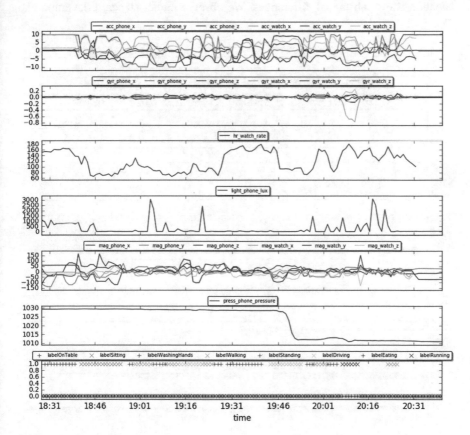

**Fig. 2.1**  Processed CrowdSignals data ($\Delta t = 60$ s)

data. To be more specific, we know that walking should provide us with some periodic changes in the accelerometer data (usually with a frequency in the order of 1 Hz) but this information is lost as a result of the aggregation. If we consider a more fine grained dataset with $\Delta t = 0.25$ s, i.e. four instances per second, we are likely to capture the stepping motion. The result is shown in Fig. 2.2 and contains a total of 31838 data points. Indeed we see a lot more variance in this data. Previously, we had just aggregated too much and lost the fine details in our dataset that might be of great value. The choice of $\Delta t$ highly depends on the task. For example, if you want to determine the step frequency of a person, your $\Delta t$ should be significantly smaller than the corresponding step period. On the other, if you want to learn about the motion state of a person, e.g. walking or sitting, $\Delta t = 1$ minute might not only be sufficient but also optimal with respect to the predictive capabilities of a model based on the aggregated data.

We have created some summary statistics of the two datasets with different $\Delta t$ in Table 2.5 to signify the differences. In addition, Fig. 2.3 shows the differences of the accelerometer data in a boxplot. We see that the extreme values and standard deviation show substantial differences. We observe higher standard deviation and

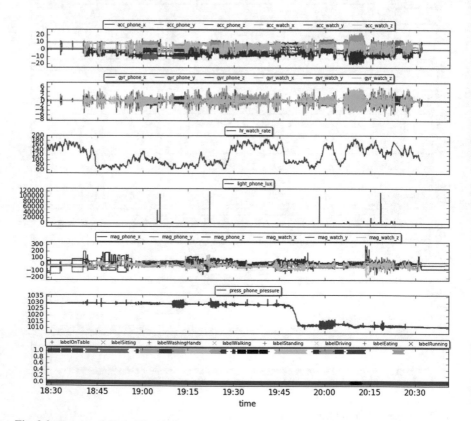

**Fig. 2.2** Processed CrowdSignals data ($\Delta t = 0.25$ s)

**Table 2.5** Statistics of processed dataset (first number listed is for $\Delta t = 60$ s, second value for $\Delta t = 0.25$ s)

*Numerical*

| Attribute | Missing (%) | Mean | | Standard deviation | | Minimum | | Maximum | |
| --- | --- | --- | --- | --- | --- | --- | --- | --- | --- |
| acc_phone_x | 0.0 | 1.1 | 1.1 | 4.2 | 4.7 | −9.1 | −11.8 | 9.0 | 17.1 |
| acc_phone_y | 0.0 | −0.9 | −0.9 | 5.6 | 6.4 | −10.1 | −15.6 | 9.8 | 14.9 |
| acc_phone_z | 0.0 | 2.0 | 2.0 | 4.7 | 5.4 | −5.3 | −11.3 | 9.6 | 11.4 |
| acc_watch_x | 7.5 | 2.0 | 2.1 | 4.9 | 5.8 | −5.8 | −12.2 | 9.6 | 22.9 |
| acc_watch_y | 7.5 | −5.2 | −5.2 | 2.4 | 3.5 | −9.1 | −20.6 | 0.2 | 10.0 |
| acc_watch_z | 7.5 | 3.6 | 3.6 | 2.7 | 4.0 | −3.4 | −12.6 | 9.2 | 13.7 |
| gyr_phone_x | 0.0 | 0.0 | 0.0 | 0.0 | 0.6 | −0.1 | −4.0 | 0.1 | 5.7 |
| gyr_phone_y | 0.0 | 0.0 | 0.0 | 0.0 | 0.4 | −0.1 | −5.0 | 0.2 | 6.5 |
| gyr_phone_z | 0.0 | 0.0 | 0.0 | 0.0 | 0.5 | −0.2 | −5.4 | 0.1 | 5.9 |
| gyr_watch_x | 8.3 | 0.0 | 0.0 | 0.1 | 0.7 | −0.8 | −6.7 | 0.1 | 6.3 |
| gyr_watch_y | 8.3 | 0.0 | 0.0 | 0.0 | 0.6 | −0.1 | −5.5 | 0.1 | 5.0 |
| gyr_watch_z | 8.3 | 0.0 | 0.0 | 0.0 | 0.8 | −0.1 | −7.0 | 0.3 | 5.5 |
| hr_watch_rate | 7.5 | 119.2 | 121.0 | 35.5 | 35.2 | 65.4 | 58.0 | 180.7 | 188.0 |
| light_phone_lux | 0.0 | 278.34 | 281.5 | 596.3 | 2220.9 | 0.0 | 0.0 | 3109.3 | 118985.0 |
| mag_phone_x | 0.0 | −13.7 | −13.5 | 46.9 | 50.6 | −121.8 | −156.4 | 115.5 | 126.6 |
| mag_phone_y | 0.0 | −3.7 | −3.8 | 44.9 | 47.9 | −139.7 | −165.4 | 80.7 | 96.8 |

(continued)

**Table 2.5** (continued)

*Numerical*

| Attribute | Missing (%) | | Mean | | Standard deviation | | Minimum | | Maximum | |
|---|---|---|---|---|---|---|---|---|---|---|
| mag_phone_z | 0.0 | 0.0 | 7.5 | 7.6 | 35.2 | 40.0 | −61.2 | −106.4 | 164.1 | 198.0 |
| mag_watch_x | 8.3 | 8.9 | −9.2 | −9.1 | 17.7 | 26.1 | −66.0 | −138.0 | 31.7 | 122.8 |
| mag_watch_y | 8.3 | 8.9 | 27.2 | 27.3 | 29.7 | 39.6 | −47.6 | −151.3 | 163.6 | 297.4 |
| mag_watch_z | 8.3 | 8.9 | −20.0 | −20.0 | 24.2 | 31.6 | −130.3 | −186.7 | 51.4 | 149.7 |
| press_phone _pressure | 0.0 | 10.3 | 1022.3 | 1022.3 | 8.3 | 8.3 | 1011.0 | 1008.6 | 1029.4 | 1033.5 |

*Categorical*

| Attribute | Value | Percentage of cases (%) | |
|---|---|---|---|
| label | OnTable | 9.0 | 7.8 |
| label | Sitting | 10.5 | 8.6 |
| label | WashingHands | 3.8 | 2.0 |
| label | Walking | 18.8 | 14.7 |
| label | Standing | 10.5 | 7.3 |
| label | Driving | 14.3 | 12.4 |
| label | Eating | 8.3 | 6.8 |
| label | Running | 4.5 | 3.8 |

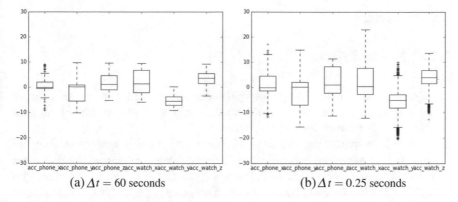

(a) $\Delta t = 60$ seconds          (b) $\Delta t = 0.25$ seconds

**Fig. 2.3** Boxplots of all accelerometer data

more extreme values for the more fine grained dataset, which is to be expected given our averaging approach to compute the values for a specific discrete time step. This is also reflected in the percentage of data points associated with each of the labels. In terms of missing values we do not see many differences for the numerical values, except for the heart rate. It seems that the sampling rate of the heart rate values is lower than the level of granularity. We will see in the next chapter how we can handle these missing values. Based on the insights we have just gained, we select the most fine grained dataset for the remainder of this book as we feel that we would lose too much information and also valuable training data if we were to use the coarse-grained variant.

## 2.4 Machine Learning Tasks

Given that we have defined and created our dataset now, we should also define some goals we want to achieve with the application of machine learning techniques to the above dataset. In general, we can set goals in sync with the different learning approaches we have briefly discussed in Sect. 1.3.2. Focusing on supervised learning we define two tasks: (1) a classification problem, namely predicting the label (i.e. activity) based on the sensors, and (2) a regression problem, namely predicting the heart rate based on the other sensory values and the activity. In the rest of the book we will see how accurate we can perform these two tasks with our dataset.

## 2.5  Exercises

### 2.5.1  Pen and Paper

1. When we measure data using sensory devices across multiple users we often see substantial differences between the sensory values we obtain. Identify at least three potential causes for these differences.
2. We have seen that we can make trade-offs in terms of the granularity at which we consider the measurements in our basic dataset. We have shown the difference between a granularity of $\Delta t = 0.25$ s and $\Delta t = 60$ s. We arrived at a choice for $\Delta t = 0.25$ s for our case, but let us think a bit more general: think of four criteria that play a role in deciding on the granularity for the measurements of a dataset.
3. We have identified two tasks we are going to tackle for the crowdsignals data. Think of at least two other machine learning tasks that could be performed on the crowdsignals dataset and argue why they could be relevant to support a user (when doing so, keep in mind the different learning approaches discussed in Sect. 1.3.2).

### 2.5.2  Coding

1. Create your own dataset for the quantified self by using your smartphone. You can create the dataset using measurement apps on your smartphone (e.g. at the time of writing Funf, SensorLog, phybox, or SensorKinetics) or other devices. Include repeated periods with different activities (please incorporate some we have seen in the crowdsignal data and some that are different) and study the variation you see in the sensory values. Be sure to include periods without any specific activities to study the background noise of the sensors. Log the intervals at which you performed the different activities.

   a. Plot and describe the data you obtain using the libraries provided with the book.
   b. Try different values for $\Delta t$ and describe the differences you see.

2. Compare the sensory values you have obtained with your measurements to those in the crowdsignals dataset over comparable activities. What would be the best way to compare the values given that the values might result from different sensors with different scales? And how different are the two datasets?
3. Find a dataset on the web that covers data from multiple users (for a list of data sources check the book's website). Note, that there are quite a few datasets that come along with an accompanying scientific article, see for example Anguita et al. [7], Banos et al. [10], or Zhang et al. [131]. Study and describe the variation you see in terms of sensory values over different users. Plot some differences that stand out and identify potential causes for these differences (e.g. by considering the ones you listed under the *pen and paper* exercises).

# Chapter 3
# Handling Noise and Missing Values in Sensory Data

In the previous chapter we have aggregated the sensory data and put it neatly into a matrix $\mathbf{X}$. By doing so, we are able to reduce some noise. However, it is likely that $\mathbf{X}$ still contains faulty or noisy measurements that pollute our data and hinder us from working on the machine learning tasks defined in Sect. 2.4. For instance, GPS sensors might be imprecise and the estimated position might jump between the northern and southern hemisphere. The same holds for accelerometers and nearly all types of sensors. Furthermore, some measurements could be missing, e.g. the heart rate monitor might temporarily fail. Although a variety of machine learning techniques exist that are reasonably robust against such noise, the importance of handling these issues is recognized in various research papers (see e.g. [128]). We have three types of approaches at our disposal that can assist us here:

1. We can use approaches that detect and remove *outliers* from the data.
2. We can *impute* missing values in our data (that could also have been outliers that were removed).
3. We can *transform* our data in order to identify the most important parts of it.

We will consider a number of approaches that fall within these categories. An overview of them is shown in Table 3.1 including some characteristics and a very brief summary. Note that nearly all these approaches are tailored towards numerical attributes, except for the distance based outlier detection algorithms and the mode and model-based imputation.

© Springer International Publishing AG 2018
M. Hoogendoorn and B. Funk, *Machine Learning for the Quantified Self*,
Cognitive Systems Monographs 35, https://doi.org/10.1007/978-3-319-66308-1_3

**Table 3.1** Methods discussed in this chapter

| Approach | Purpose | Specific for $\mathbf{X}^{\mathcal{T}}$ | Number of attributes considered | Brief summary |
|---|---|---|---|---|
| Chauvenets criterion | Outlier detection | No | 1 | Identify values for an attribute that are unlikely given a single normal distribution to describe the data. |
| Mixture model-based outlier detection | Outlier detection | No | 1 | Identify values for an attribute that are unlikely given a combinations of distributions to describe the data. |
| Simple distance-based outlier detection | Outlier detection | No | $1, \ldots, p$ | Identify instances with a great distance to other points. |
| Local outlier factor | Outlier detection | No | $1, \ldots, p$ | Identify instances that have a lower local density than its neighboring points. |
| Mean imputation | Missing value imputation | No | 1 | Impute the mean value for an attribute for an unknown value or outlier. |
| Median imputation | Missing value imputation | No | 1 | Impute the median value for an attribute for an unknown value or outlier. |
| Mode imputation | Missing value imputation | No | 1 | Impute the mode value for an attribute for an unknown value or outlier. |
| Interpolation-based imputation | Missing value imputation | Yes | 1 | Impute the value for an attribute by extrapolating the previous and next measurement. |
| Model-based imputation | Missing value imputation | No | 1 | Impute the value for an attribute by creating a model to predict it. |
| Kalman filter | Outlier detection & Missing value imputation | Yes | $1, \ldots, p$ | Estimate expected values based on historical observations and impute them when values are too deviant. |
| Lowpass Butterworth filter | Transformation | Yes | 1 | Remove periodic irrelevant data of a single attribute over time. |
| Principal Component Analysis | Transformation | No | $p$ | Condense most of the variability of the data in a set of new features. |

# 3.1 Detecting Outliers

When we get started with our dataset we are potentially confronted with some extreme values that are highly unlikely to occur. We will call these *outliers*. When working with data from physical sensors as in our case, outliers are very likely. We define an outlier as follows:

**Definition 3.1** An outlier is an observation point that is distant from other observations (cf. [53]).

Observations in the sense of this definition can be two different things: we can consider single values of one attribute $X_j$ as an observation $(x_i^j)$, or we can consider complete instances as an observation $(x_i)$. We will see that some approaches we discuss can only handle single attributes while others can cope with complete instances.

We can have two types of outliers: those caused by a *measurement error* and those simply caused by *variability* of the phenomenon that we observe or measure. Typically, we are interested in getting a full picture on what was caused by the phenomenon under study while we try to get rid of the measurement errors. When considering our example Arnold, a measurement of a heart rate of 300 would be considered a measurement error (unless our friend has some form of superpowers) whereas a heart rate of 195 might be very uncommon but could simply be a measurement of Arnold trying to push his limits. While we would clearly like to remove the measurement errors or replace them with more realistic values (which will hopefully yield a better performance of our machine learning approaches), we should be very careful not to remove the outliers caused by the variability in the measured quantity itself. Obviously, life will not always be as clear cut, and we are in need of approaches that can assist us in the process. One approach is to remove measurement errors based on domain knowledge rather than based on machine learning. For example, we know that a heart rate can never be higher than 220 beats per minute and cannot be below 27 beats per minute (the current world record). So, we remove all values outside of this range and interpret them as missing values. This will often be the right choice, but there are situations in which outliers by their existence carry information, e.g. a heart rate above 220 bpm is not possible but might reflect a situation of extreme physical stress causing the chest strap not to work properly. Hence, there is the possibility that we filter out important information.

An additional problem we might encounter is that domain knowledge is not widely accessible or to a large extent it is unknown how to define outlier for a domain. What can we do, if we do not possess this type of domain knowledge and have no up-front knowledge on what an outlier is? Below, we will treat various approaches that can help us to remove outliers. We will assume we do not have any knowledge on what outliers are. Hence, we consider it being an unsupervised problem. Be aware that this process is **dangerous** as there is a high risk of removing points that are not measurements errors and might in fact be the most interesting points in our dataset. One thing we can do is perform visual inspection to make sure we do not remove any valuable information. Alternatively, we can also just try whether we improve on our machine learning tasks when we remove them. We roughly follow the categorization of [59] for outlier detection algorithms, discussing distribution-based models first, followed by distance-based approaches.

### 3.1.1  Distribution-Based Models

The first approach we will consider for outlier removal is based on the probability distribution of the data. Here, the data should follow a certain known distribution and we remove those that are outside of certain bounds of the distribution. These approaches are mainly targeted at single attributes $X_j$.

#### 3.1.1.1  Chauvenets Criterion

When we consider Chauvenets criterion (cf. [35]), we assume the data to follow the normal distribution. Given that we have $N$ measurements for attribute $X_j$, we compute the mean $\mu$ and standard deviation $\sigma$ of our data:

$$\mu = \frac{\sum_{n=1}^{N} x_n^j}{N} \tag{3.1}$$

$$\sigma = \sqrt{\frac{\sum_{n=1}^{N} (x_n^j - \mu)^2}{N}} \tag{3.2}$$

Together, these values define a normal distribution $\mathcal{N}(\mu, \sigma^2)$. According to Chauvenet's criterion we reject a measurement from a dataset of size $N$ when its probability of observation is less than $\frac{1}{2N}$. A generalization of this criterion is to replace the value 2 with a parameter $c$, we will follow this generalization in the remainder of this explanation. We can compute the probability of observing a value of at most $x_i^j$ as follows:

$$P(X \leq x_i^j) = \int\limits_{-\infty}^{x_i^j} \frac{1}{\sqrt{2\sigma^2 \pi}} e^{-\frac{(u-\mu)^2}{2\sigma^2}} \delta u \tag{3.3}$$

A point is considered an outlier when one of the following two cases holds:

**Fig. 3.1** Outlier detection based on Chauvenet's criterion

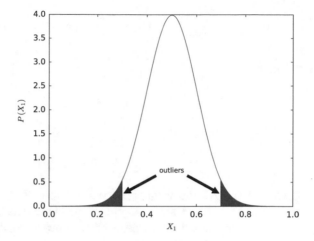

$$(1 - P(X \le x_i^j)) < \frac{1}{cN} \qquad (3.4)$$

$$P(X \le x_i^j) < \frac{1}{cN} \qquad (3.5)$$

where $c$ is a positive constant number roughly between 1 and 10 that specifies the degree of certainty for the identification of outliers given the assumption of a normal distribution. A higher $c$ corresponds to higher chance that identified outliers are truly outliers. Graphically, the outliers are visualized in Fig. 3.1. The red areas reflect a low probability (less than $\frac{1}{cN}$) of observing measurements that are not outliers—in other words we assume measurements in this area to be outliers (remember, we do not have ground truth here, although we do assume the normal distribution). Alternatives to Chauvenet's criterion exist that are based on the same assumptions yet a bit more sophisticated, see e.g. [52].

### 3.1.1.2 Mixture Models

While the previous approach is straightforward, it does assume that a single distribution can be fitted to our measurements. This might not always be realistic. If we think of accelerometer data, our data could for example follow a combination of two normal distributions one bump for measurements in case of a user being inactive and one for the same user being active. We can solve this problem with *mixture models*. Assuming we have $K$ distributions to describe our data, e.g. $K$ normal distributions $\{\mathcal{N}(\mu_1, \sigma_1), \ldots, \mathcal{N}(\mu_K, \sigma_K)\}$, we would then like to find the values of the parameters for those individual distributions (i.e. $\mu_1 \ldots \mu_k, \sigma_1, \ldots, \sigma_K$) that when combined best describe the data. We formulate the probability of observing a value $x_n^j$ for a specific measurement:

$$p(x_n^j) = \sum_{k=1}^{K} \pi_k \mathcal{N}(x_n^j | \mu_k, \sigma_k) \tag{3.6}$$

With

$$\sum_{k=1}^{K} \pi_k = 1 \tag{3.7}$$

$$\forall k : 0 < \pi_k \leq 1 \tag{3.8}$$

Here, $\pi_k$ expresses the weight of each distribution. The sum of the weights is scaled to 1 to make sure the total area under the curve remains 1. We need to find the parameters that maximize the probability of observing the data we have measured, specified by means of the likelihood:

$$L = \prod_{n=1}^{N} p(x_n^j) \tag{3.9}$$

In other words, we maximize the product of the probabilities of observing our attribute values; the higher the probabilities of the individual attribute values the higher the product. One way to do so is the Expectation-Maximization algorithm (cf. [18]) which you can explore in the exercises. Once we have found the best parameters, we can consider identifying outliers again by considering the probability of each observation. Points with the lowest probabilities are candidates for removal. The exact criterion to apply here depends on the data at hand. In order to find the best number of distributions to fit the data, multiple approaches have been developed, see [61] for an overview.

## 3.1.2  Distance-Based Models

A second type of algorithm to detect outliers is to consider the distance between a point and the other points in the dataset. We will treat distance metrics between points in more detail in Chap. 5. For now assume that we have a metric to compute the distance between two instances $x_i$ and $x_j$ called $d(x_i, x_j)$. This is different from the distribution-based approach which only focused on individual attributes. Of course, one can also consider individual attributes for the distance-based approaches (just consider $p = 1$).

### 3.1.2.1 Simple Distance-Based Approach

The first approach takes a global view towards the data: we consider the distance of a point to all other points. We define a certain minimum distance $d_{min}$ within which we consider a point to be close to another point. We say that a point is an outlier when more than a fraction $f_{min}$ of the points in the dataset is at a distance of more than $d_{min}$ (cf. [72]) from that point. Formally:

$$outlier(x_i) = \begin{cases} 1 & \frac{\sum_{n=1}^{N} d\_over(x_i, x_n, d_{min})}{N} > f_{min} \\ 0 & otherwise \end{cases} \tag{3.10}$$

where

$$d\_over(x, y, d_{min}) = \begin{cases} 1 & d(x, y) > d_{min} \\ 0 & otherwise \end{cases} \tag{3.11}$$

Again, the parameter settings for $f_{min}$ and $d_{min}$ are crucial for this approach in order to work well. An example is shown in Fig. 3.2, where we have two relevant attributes, $X_1$ and $X_2$ representing the x- and y-axis of the accelerometer data of Arnold. The red dots represent data points where the x-axis and y-axis stand for their values of $X_1$ and $X_2$ respectively. Assuming we want to determine whether the black dot is an outlier, we compute for each point whether it occurs within distance $d_{min}$. This distance from our point is indicated by the circle, assuming a certain distance metric.

**Fig. 3.2** Outliers based on distance

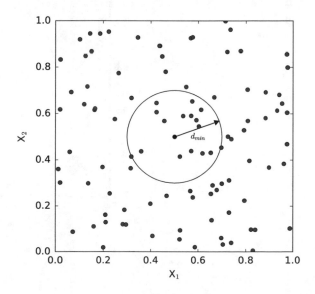

We see that fifteen other points lie within this distance while all other are outside. Depending on our parameter settings we could consider this an outlier or not.

### 3.1.2.2   Local Outlier Factor

Instead of taking a global look at points the local outlier factor approach (cf. [30]) only takes points into account that surround it. Some areas in our data space might be quite dense while others are not. Taking this into account might improve our detection of outliers. In addition, the approach specifies a degree of outlierness (i.e. the likelihood of an instance being an outlier). This was missing in the previous distance based approach, although it would not be difficult to change the previous approach slightly to account for this. Imagine the scenario shown in Fig. 3.3. We see data points on the lower left that form a sort of cluster, let us call this cluster 1. On the upper right we see a similar, yet less dense cluster, which we call cluster 2. Consider the two points visualized by black dots, point 1 being the lower left point, point 2 the one in the upper right. Given the shape of cluster 1, point 1 is likely to be considered an outlier while point 2 is probably not, given the shape of cluster 2. Hence, even though the distances to the points in the cluster are similar we treat them differently.

Let us dive into the approach in a bit more detail. The first step taken is to define the k-distance $k_{dist}$ of a point $x_i$. This is defined as the largest distance among the distances of the $k$ closest points. To describe this, we define the following constraints for $k_{dist}(x_i)$ of a point $x_i$:

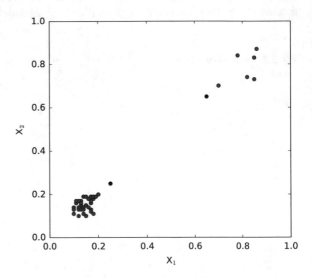

**Fig. 3.3**  Example outliers for local outlier factor

$$|\{x|x \in \{x_1, \ldots, x_{i-1}, x_{i+1}, \ldots, x_N\} \wedge d(x, x_i) \leq k_{dist}(x_i)\}| \geq k \quad (3.12)$$
$$|\{x|x \in \{x_1, \ldots, x_{i-1}, x_{i+1}, \ldots, x_N\} \wedge d(x, x_i) < k_{dist}(x_i)\}| \leq (k-1) \quad (3.13)$$

In other words, there should be $k - 1$ points or less with a distance less than $k_{dist}$ and at least one point with the same distance. The set of neighbors within this distance is called $k_{dist\_nh}$:

$$k_{dist\_nh}(x_i) = \{x|x \in \{x_1, \ldots, x_{i-1}, x_{i+1}, \ldots, x_N\} \wedge d(x, x_i) \leq k_{dist}(x_i)\} \quad (3.14)$$

We define the reachability distance of a point $x_i$ to another point $x$ as:

$$k_{reach\_dist}(x_i, x) = max(k_{dist}(x), d(x, x_i)) \quad (3.15)$$

This expresses that a reachability distance is the real distance if the point $x_i$ is not among the $k$ nearest points of $x$ (in that case the value for $d(x, x_i)$ will be larger than $k_{dist}(x)$) and otherwise it is $k_{dist}$ of that point, so we set the distance value of all points within $k_{dist}(x)$ equal to $k_{dist}(x)$. Next, consider the local reachability density around our point $x_i$:

$$k_{lrd}(x_i) = 1 / \left( \frac{\sum_{x \in k_{dist\_nh}(x_i)} k_{reach\_dist}(x_i, x)}{|k_{dist\_nh}(x_i)|} \right) \quad (3.16)$$

In our definition, we look at the neighbors $x$ of $x_i$ within $k_{dist}$, their reachability distance to $x_i$, and divide this by the number of neighbors. Intuitively this says something on how close point $x_i$ is to its neighbors. We divide 1 by this number, so the lower the average distance to ones neighbors, the higher the value. In order to see how much of an outlier the point is compared to its neighbors, we consider the local reachability density of those neighbors and compare them:

$$k_{lof}(x_i) = \frac{\sum_{x \in k_{dist\_nh}(x_i)} \frac{k_{lrd}(x)}{k_{lrd}(x_i)}}{|k_{dist\_nh}(x_i)|} \quad (3.17)$$

We compare the values for our point $x_i$ and our neighboring points. Remember that a high value for $k_{lrd}$ represents a closer proximity to neighbors. This formula expresses that the higher the scores for $x_i$ on the $k_{lrd}$ compared to its neighbors, the lower the local outlier factor will become, which makes perfect sense.

## 3.2 Imputation of Missing Values

Obviously, our dataset could contain a lot of missing values. This could be caused by a lot of outliers which were removed, or possibly by sensors not providing information at certain points in time. There are different ways to replace these missing values, we refer to this process as *imputation*.

The first approach we can take is to impute the *mean* value of an attribute calculated over the instances where the value is known. This is a common approach. The approach does have disadvantages when we have data with a lot of extreme values, as this severely impacts the value of the mean. The *median* is a robust alternative for these cases as it is less sensitive to extreme values. Note that all these approaches target numerical values. For categorical values, we can use the *mode*.

A more sophisticated approach is to predict the missing value for attribute $j$ of instance $i$ ($x_i^j$) using statistical models such as linear regression. In general, there are two ways this can be done:

1. we consider the values for the other attributes of the same instance and predict $x_i^j$ using those values (i.e. $x_i^1, \ldots, x_i^{j-1}, x_i^{j+1}, \ldots, x_i^p \rightarrow x_i^j$);
2. we take the previous (and possibly next) values of the same attribute. For the latter we need a temporal dataset ($\mathbf{X}^{\mathcal{T}}$). Hence, we predict in the following way: $x_1^j, \ldots, x_{i-1}^j, x_{i+1}^j, \ldots, x_N^j \rightarrow x_i^j$.

We do not need to use the entire set of attributes (first case) or instances (second case) but can also consider a subset. A simple example of the second approach is to take the previous and next value of the specific attribute and average the values (again assuming numerical values), i.e.

$$x_i^j = \frac{x_{i-1}^j + x_{i+1}^j}{2} \tag{3.18}$$

Or in case we know that measurement $x_{i-k}^j$ is the last available measurement while $x_{i+l}^j$ is the first next measurement we can compute it in the following way (assuming a fixed sampling rate of our data):

$$x_i^j = x_{i-k}^j + k \cdot \frac{x_{i+l}^j - x_{i-k}^j}{(k+l)} \tag{3.19}$$

This is a simple form of linear interpolation and works under the assumption that $x_i^j$ follows a linear trend. Of course, this does not work for the first and last time points. In these cases we can either use the next (or previous) couple of time point and extrapolate the trend from those points. The method described above belongs to the set of parametric imputation methods that rely on making assumptions on the distributions of the data as well as the underlying regression relationships. Instead, it is also possible to use non-parametric approaches to imputation. We can think of these methods as fitting locally-weighted regressions to the data. That is why these approaches are also called local imputation schemes [3].

Some approaches allow for the usage of values of multiple attributes over time. An example of this is the Kalman filter which we will discuss in the next section.

## 3.3 A Combined Approach: The Kalman Filter

An approach that identifies outliers and also replaces these with new values is the Kalman filter (cf. [66]). The Kalman filter provides a model for the expected values based on historical data and estimates how noisy a new measurement is by comparing the observed values with the predicted values. Imagine Arnold running through his favorite park in Amsterdam, the Vondelpark. We continuously obtain GPS signals on his whereabouts. Suddenly we see a strange measurement: Arnold is supposed to have moved one kilometer in 10 seconds. While we should obviously never underestimate Arnold's physical shape we also know he is neither superman nor a Ferrari. The Kalman filter can find this strange anomaly and will insert a more reliable value instead. This can also work really well in a real time setting. Again, we need a dataset with temporal ordering for this to work.

Let us dive into the method in a bit more detail. In the Kalman filter, we distinguish between a latent state $s_t$ and the measurements that can be performed based on the state, in our case $x_t$, or $x_t^j$ if we want to consider single measurements. In our example case, the state of Arnold would be the actual presence at a certain location, his actual physical state, etcetera. Our measurement would be the GPS coordinate, activity level, heart rate and so on. We can express the next value of a state as:

$$s_t = F_t s_{t-1} + B_t u_t + w_t \tag{3.20}$$

Here, $F_t$ and $B_t$ are matrices while $s_t$ is a vector that represents the previous value of the state, $u_t$ is the control input to the state (we might want to adjust the state, e.g. by sending a message to Arnold, we will return to this in Chap. 9), and $w_t$ represents process white noise. $F_t$ expresses how the previous state influences the new state (by means of weights associated for each component of the state) while $B_t$ represents how the control input influences the different components of the next state.

The value for the measurement associated with the state is:

$$x_t = H_t s_t + v_t \tag{3.21}$$

$H_t$ is again a matrix while $v_t$ is the measurement white noise. The noise values are assumed to be taken from a multivariate normal distribution with covariance matrix $Q_t$ (representing the process noise covariance) and $R_t$ (the measurement noise covariance) respectively:

$$w_t = \mathcal{N}(0, Q_t) \tag{3.22}$$
$$v_t = \mathcal{N}(0, R_t) \tag{3.23}$$

Given this model, we can start to make predictions of the next state. Let $\hat{s}_{t|t-1}$ represent the estimation of the state value $s_t$ given observations up to $t-1$ (i.e. an a priori estimate) and $\hat{s}_{t|t}$ the a posteriori estimate given all observations including the observation at time $t$.

Remember that the noise plays an important role, since it determines the uncertainty we encounter when estimating the state. We assume that we have a matrix containing the level of noise or error we expect to see, denoted as $P_t$. The matrix contains the covariance and variances of the Gaussian probability density functions that characterize the error in our predictions. $P_{t|t-1}$ represents our a priori estimation of the error while $P_{t|t}$ is our a posteriori estimation. We start by making our predictions:

$$\hat{s}_{t|t-1} = F_t \hat{s}_{t-1|t-1} + B_t u_t \tag{3.24}$$

$$P_{t|t-1} = \mathbb{E}[(s_t - \hat{s}_{t|t-1})(s_t - \hat{s}_{t|t-1})^T] \tag{3.25}$$

Equation 3.24 is easy to understand given the previous Eq. 3.20 for the next value of the state. Equation 3.25 updates our estimation on the error based on our estimation of the state. Note that we cannot directly compute $P_{t|t-1}$ this way as we do not know $s_t$, but you can derive the value of $P_{t|t-1}$ in a different way. It is beyond the scope of this book to go into detail on this. The equation above does provide some intuition on the meaning of $P_{t|t-1}$.

The next step is to update our model based on our a priori predictions and the actual observations we perform. Here our matrix $H_t$ plays an important role, recall that this is the mapping from a state to measurements associated with that state. First, we can determine the difference between a measurement $x_t$ and our a priori estimate for the measurement, we will refer to this as $e_t$:

$$e_t = x_t - H^T \hat{s}_{t|t-1} \tag{3.26}$$

And we can create an estimation for the covariance for our measurements $S_t$ given our matrix $H_t$:

$$S_t = H_t P_{t-1|t-1} H_t^T + R_t \tag{3.27}$$

In order to make the best estimate of the state given the error we observe between the actual and expected measurements we can define the *optimal Kalman gain*, defined as:

$$K_t = P_{t|t-1} H_t^T S_t^{-1} \qquad (3.28)$$

And we use this to define the a posteriori estimation of the state based on our a priori estimate and the error between our model and the actual measurements:

$$\hat{s}_{t|t} = \hat{s}_{t|t-1} + K_t e_t \qquad (3.29)$$

Finally, we update our covariance matrix:

$$P_{t|t} = (I - K_t H_t) P_{t|t-1} \qquad (3.30)$$

The sequence of equations might be a bit overwhelming, but we hope that the intuition behind the approach has become clear: Basically, we distinguish between a state, that is not directly observed, and a measurement. Kalman filtering is then a set of transition equations that uses measurements to describe how the state is evolving. There are ample books and tutorials that discuss Kalman filtering in more detail [130].

We can use the Kalman filter by letting it run over our data, and then analyze when the deviation (i.e. $e_t$) is too big. In case it is (apparently it is noise), we can use our estimate and otherwise we use the original value. Of course, we keep on updating our model independent of whether we accept a measurement or not since this allows us to better estimate the noise levels.

Next to using the Kalman filter to detect and impute outliers, it can also be used for sensor fusion to extract information from a combination of sensors. Think again of our friend Arnold we could fuse data from various sensors (e.g. GPS location, accelerometer, step counter, speed) to determine his exact location, being our latent state. See [111] for an example of how to use the Kalman filter for sensor fusion.

## 3.4   Transformation

The next approaches for handling noisy data is to transform our data in a way that subtle noise (not the huge outliers we have seen before) is filtered and the parts of our data that explain most of the variance are identified. We will explain two approaches: the lowpass filter (which can be applied to individual attributes) and Principal Component Analysis, which works across the entire dataset.

### 3.4.1  Lowpass Filter

The lowpass filter can be applied to data that is of temporal nature (i.e. $\mathbf{X}^{\mathcal{T}}$), and assumes that there is a form of periodicity. Think about accelerometer data for example, if we are walking, we will see periodic measurements in our accelerometer data at a frequency around 1 Hz as walking is a repetitive pattern. When we process our data, we can filter out such periodic constituents based upon their frequency. We could say for instance that any measurement we perform that is at a higher frequency than our walking behavior is irrelevant and can be considered as noise (they might actually hamper our machine learning process). Hence, we want to remove the data originating from a form of periodicity above a certain frequency, and leave the data at lower frequencies untouched. How we move from measurement values over time to frequencies is explained in more detail in Chap. 4 when we discuss Fourier Transformations. An example for cutting out parts of the frequency spectrum and the corresponding periodic behavior is the Butterworth filter. Assuming $f_c$ to be the cutoff frequency we represent the transfer function $G$ (usually specified in dB due to the traditional application in the audio domain) of the signal with frequency $f$ by the following (simplified) equation:

$$|G(f)|^2 = \frac{1}{1 + (f/f_c)^{2n}} \tag{3.31}$$

We see that the higher the frequency $f$, the lower the magnitude of the transfer function $G$, which is exactly what we want: we want to filter out the high frequency data and let the low frequency data pass. The parameter $n$ is the order of the filter. The higher the order, the more steeply the magnitude of the frequencies above the cutoff frequency $f_c$ drop. Let us consider an application of the filter to exemplify its working.

Figure 3.4 shows an example that combines two sinusoid functions with different frequencies, one with a frequency of 1 Hz and one with a much lower frequency, namely 0.1 Hz. We apply a Butterworth filter with a cutoff frequency of 0.5 Hz (i.e. everything with a higher frequency is filtered) to the combined data and see that we obtain the single low frequency sinusoid: we have filtered out the high frequency data. Lowpass filters have for instance been used in [7, 32, 36, 128] to clean up the data.

### 3.4.2  Principal Component Analysis

The Butterworth filter addresses individual attributes, transforms the signal into the frequency domain, and then picks a specific part of the frequency spectrum (e.g. low frequencies). We can also consider a set of attributes at the same time and extract features that explain most of the variation observed over all attributes. To

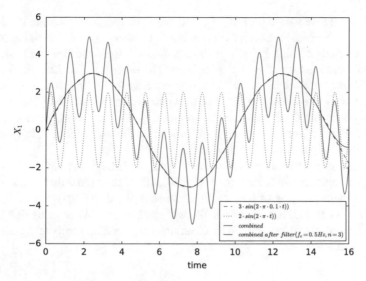

**Fig. 3.4** Example application of Butterworth filter

**Fig. 3.5** Example dataset
for principal component
analysis

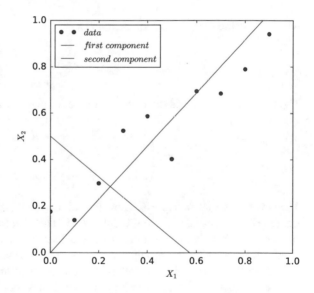

understand this, let us consider Fig. 3.5, in which we have two attributes $X_1$ and
$X_2$. The measurements might represent the accelerometer data of Arnold and are
expressed by the red points. We see that an increase in the measurement value for
$X_1$ (acceleration on the x-axis) goes hand in hand with an increasing value for $X_2$
(acceleration on the y-axis).

How can we explain the variance in the data? We can actually see that the solid red
line describes our data in a reasonably precise way. If we would know the equation

of this line, and we would project all points onto that line (so we ignore the distance to the line) we would be able to express our data points by a single value $X_{new}$ on the line instead of value pairs. In fact, the distance from the line could be noise, which we could get rid of in one go as well. The other line (the blue line) is a line perpendicular (at an angle of $90°$) to our previously found line. It can be used as a secondary axis to express the distance of a point from our previous line. If we use both, we obviously do not lose any information, and hence, we might not get rid of the noise. Here, we only exemplify this procedure for two attributes, however, it is applicable to an arbitrary number of attributes.

The main goal of Principal Component Analysis (PCA) (see e.g. [64]) is to find vectors that represent these lines (or hyperplanes if we have more than two attributes) and order them in terms of how much variance in the data is explained. There are different ways to find these vectors. We will explain one. We start by defining the co-variance matrix of our data. The covariance between an attribute $X_i$ and $X_j$ is defined as follows:

$$cov(X_i, X_j) = \frac{\sum_{n=1}^{N}(x_n^i - \bar{X}_i)(x_n^j - \bar{X}_j)}{N} \qquad (3.32)$$

where $\bar{X}_k$ is the mean value for attribute $X_k$:

$$\bar{X}_k = \frac{\sum_{n=1}^{N} x_n^k}{N} \qquad (3.33)$$

The covariance matrix is then defined in the following way:

$$C = \begin{pmatrix} cov(X_1, X_1) & \cdots & cov(X_1, X_p) \\ \vdots & \ddots & \vdots \\ cov(X_p, X_1) & \cdots & cov(X_p, X_p) \end{pmatrix} \qquad (3.34)$$

The covariance matrix expresses how the values of our attributes relate to each other, i.e. to what extent values are are correlated. The covariance of the same attribute (i.e. $cov(X_i, X_i)$) is the variance of the attribute. If we divide each element of the covariance matrix by the product of the per attribute standard deviations, i.e. $\sigma_i \sigma_j$, we obtain the matrix of all correlation coefficients. If an element of this matrix is 1, values of both attributes across our instances deviate from the mean in the same way (i.e. if we have a low value of one attribute compared to the mean, the same holds for the other attribute). In case of a correlation of -1 they behave in precisely the opposite way. Note that our matrix is symmetric ($cov(X_i, X_j) = cov(X_j, X_i)$).

How does this help us to find our line in Fig. 3.5 to describe the data? Well, once we have obtained this matrix we can find its *eigenvectors*. These are vectors that when they are multiplied with our covariance matrix, we end up with (a multiplication of) our original vector, i.e.

$$\begin{pmatrix} cov(X_1, X_1) & \cdots & cov(X_1, X_p) \\ \vdots & \ddots & \vdots \\ cov(X_p, X_1) & \cdots & cov(X_p, X_p) \end{pmatrix} \cdot \begin{pmatrix} v_1 \\ \vdots \\ v_p \end{pmatrix} = f \cdot \begin{pmatrix} v_1 \\ \vdots \\ v_p \end{pmatrix} \qquad (3.35)$$

Here, the factor of the multiplication $f$ is the *eigenvalue* of the *eigenvector*. To avoid counting multiplications of our eigenvectors twice we only consider the normalized eigenvectors:

$$\begin{pmatrix} sv_1 \\ \vdots \\ sv_p \end{pmatrix} = \begin{pmatrix} v_1 \\ \vdots \\ v_p \end{pmatrix} \div \sqrt{v_1^2 + \cdots + v_p^2} \qquad (3.36)$$

Typically, each covariance matrix of dimensions $p \times p$ (remember that we have $p$ attributes) has exactly $p$ normalized eigenvectors. These eigenvectors are perpendicular. To find these eigenvectors ample approaches are available which we will not discuss here. These eigenvectors in fact are precisely the lines we have considered in our problem setting before. The normalized eigenvector with the highest eigenvalue is the line (or hyperplane for $p > 2$ attributes) that explains most variance in the data. This is called the principal component (hence the name of the approach).

Let $\{sv_1^i, \ldots, sv_p^i\}$ represent the values of the eigenvector with the $i$th highest eigenvalue $f_i$. We can project our data to these new axes (just as we have explained for the example). If we select $n \le p$ eigenvectors we obtain a new (projected) dataset as follows:

$$\begin{pmatrix} x_1^1 & \cdots & x_1^p \\ \vdots & \ddots & \vdots \\ x_N^1 & \cdots & x_N^p \end{pmatrix} \cdot \begin{pmatrix} sv_1^1 & \cdots & sv_1^n \\ \vdots & \ddots & \vdots \\ sv_p^1 & \cdots & sv_p^n \end{pmatrix} = \begin{pmatrix} X_1^1 & \cdots & X_1^n \\ \vdots & \ddots & \vdots \\ X_N^1 & \cdots & X_N^n \end{pmatrix} \qquad (3.37)$$

In the matrix on the right hand side of Eq. 3.37, each row still represents an instance of our original dataset. However, we have reduced it to $n$ attributes instead of $p$.

If we take $n < p$, we do lose some information. Given our ordering however, the last eigenvectors are not likely to explain a lot of variance. Typically, the first number of components (i.e. those with the highest eigenvalues) explain most of it. In the case study we will see a typical example and illustrate that there is often a clear cut-off point.

Let us return to the previous example from Fig. 3.5. We have already visualized the principal component (solid red line) and the second component (dashed blue line). The principal component is specified as $< sv_1^1, sv_2^1 > = < 0.7044, 0.7089 >$ with $f_1 = 22.13$ and $< sv_1^2, sv_2^2 > = < -0.7089, 0.7044 >$ with $f_2 = 0.0392$. We can see that the eigenvalue of the first vector is much higher. This makes a

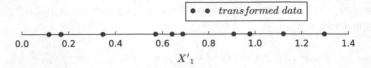

**Fig. 3.6** Our example data after application of the first principal component

**Fig. 3.7** Our example data
after application of the first
two (i.e. all) principal
components

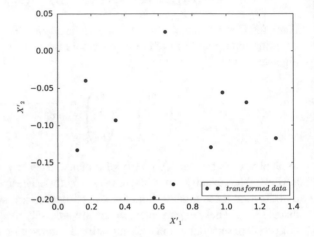

lot of sense, as it explains nearly all variance while the second one does not significantly contribute. If we apply only the first component and transform the data, we obtain the new dataset visualized in Fig. 3.6. Hence, we only have a single dimension representing the positioning of a data point on the line spanned by the first principal component. Figure 3.7 visualizes the new dataset when we use both components. Here, only the values have changed due to the two new axes. We see PCA being applied in various research papers related to the quantified self, including [16, 19].

Great, we are now able to extract useful features from our data using PCA. The big disadvantage is that we often lose the ability to interpret our models as we now have a new space with new values that are not immediately interpretable by a domain expert. In low dimensional cases it can be helpful to look for correlations between original attributes and the principal components to support interpretation.

## 3.5  Case Study

Let us consider the crowdsignals dataset again. We will now iteratively try the different approaches we have presented in this section and select the most appropriate ones to process our dataset. We will pass the various approaches in the same order they were explained before. First, we will show how to apply outlier detection in this dataset and filter extreme values. Then, we will impute missing values. Next, we will look at Kalman filtering as an integrated alternative approach to the previous steps. Finally, we end with the application of lowpass filtering and principal component analysis.

### 3.5.1 Outlier Detection

We have seen various outlier detection algorithms, but which one is best for our current dataset? To determine this, we will explore two representative types of measurements, namely the *acc_phone_x* which varies widely (similar to the other accelerometer measurements, magnetometers and gyroscope), and *light_phone_lux* which seems to be pretty stable except for a few extreme values (similar to heart rate and pressure). First, we try to explore the parameter setting of the different algorithms that result in reasonable detection of outliers. We do this by visual inspection and by seeing whether points we would visually consider to be outliers are indeed flagged without flagging points that seem normal. Figure 3.8 shows our four outlier approaches applied to *acc_phone_x* while Fig. 3.9 shows the same for *light_phone_lux*. These are the type of figures we use for our visual inspection. Note that we consider attributes in isolation here while our distance-based approaches would allow us to look at combinations of attributes. We have made this choice because we want to compare all approaches, and we still obtain good results in terms of finding outliers. To generate Figs. 3.8 and 3.9 we used the following parameter settings:

- Chauvenet's criterion: we set the value $c = 2$, according to the traditional Chauvenet criterion.
- Mixture models: we use 3 mixture components and a single iteration of the algorithm
- Simple distance-based approach: we set $d_{min} = 0.1$ and $f_{min} = 0.99$ and use Euclidean distance.
- Local outlier factor: we use 5 for the value of $k$ and Euclidean distance.

In Figs. 3.8 and 3.9 we can see that Chauvenet's criterion does signal outliers for the *light_phone_lux* attribute: we find 33 outliers that seem to make sense. For the *acc_phone_x* we do not find any outliers, and visual inspection indeed shows that there are not very clear outliers. Note that we did not explicitly check whether the distribution of the values follow the normal distribution in this case, but the visual inspection does show that the outliers found are appropriate. You can explore this issue more in the exercises. The mixture models seems to work fine for *light_phone_lux* as well: extreme and rare values get a probability of observing of around 0. For *acc_phone_x* we again do not see very clear outliers; this is a sign that very obvious outliers are indeed missing. Our simple distance-based outlier detection finds outliers for the two examples: we see some outliers for both cases (27 for *light_phone_lux* and 11 for *acc_phone_x*). Finally, the local outlier factor does show changes in values

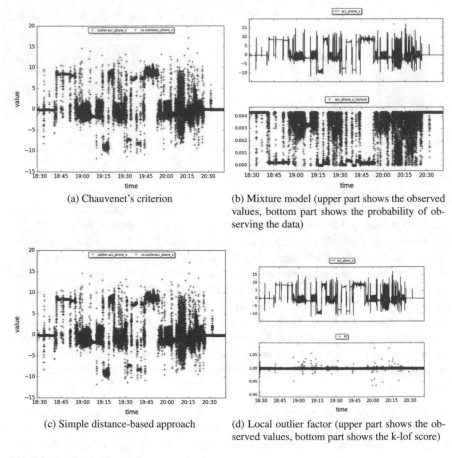

(a) Chauvenet's criterion

(b) Mixture model (upper part shows the observed values, bottom part shows the probability of observing the data)

(c) Simple distance-based approach

(d) Local outlier factor (upper part shows the observed values, bottom part shows the k-lof score)

**Fig. 3.8** Outlier for the attribute *acc_phone_x*

for outliers, but is in our opinion less clear compared to the simpler distance-based approach. In addition, it is computationally more demanding. Based on our observations, we have decided to apply a filtering of the outliers (replacing them with an unknown value) using the Chauvenet criterion: we want to be on the safe side and not throw away data points for which it is not so obvious that they are outliers. We apply this to all attributes except for the labels (that are just binary and do not contain outliers). Using a single parameter across all attributes has a severe risk which we are completely aware of, but visual inspection showed that the outliers that were removed seemed fairly reasonable.

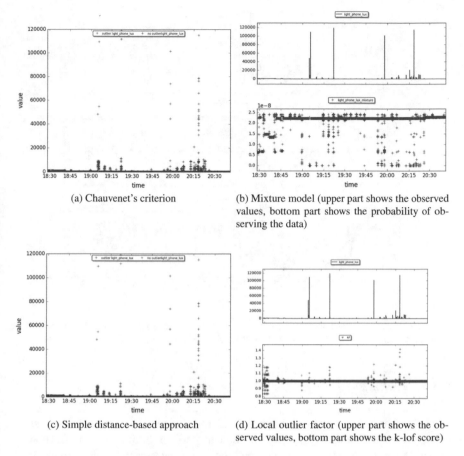

(a) Chauvenet's criterion

(b) Mixture model (upper part shows the observed values, bottom part shows the probability of observing the data)

(c) Simple distance-based approach

(d) Local outlier factor (upper part shows the observed values, bottom part shows the k-lof score)

**Fig. 3.9**  Outlier for the attribute *light_phone_lux*

## 3.5.2  Missing Value Imputation

Now that we have removed the extreme values, we are left with a number of missing values. One option is removing the instances that contain missing values. However, this would result in a loss of valuable data. Therefore, we consider imputation of the missing values. We have seen that the heart rate attribute contains most missing values, so let us use the heart rate as an example. Figure 3.10 shows the result of imputation by using the mean and interpolation. Clearly, for this type of time series, interpolation techniques are preferred for imputation: it results in much more natural values. This holds for temporal sequences in general, but if we do not have this temporal ordering it is impossible. We apply this to all attributes with missing values (except the label attributes).

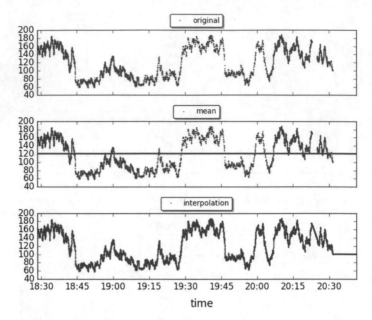

**Fig. 3.10** Missing value imputation for *hr_watch_rate*. The *top panel* shows the original data followed by imputation using the mean, and interpolation respectively

### 3.5.3   Kalman Filter

As an alternative to our two step approach (identifying outliers and imputing missing values) we can also apply the Kalman filter to perform both tasks simultaneously. However, we do not have a known model that relates our observations to true values. Therefore, we are required to use a very simple model which directly links observations to real values per attribute. Other parameters of the Kalman filter (e.g. $Q_t$, $R_t$) can be automatically tuned towards the dataset. Despite its simplicity, this approach is able to capture whether a measured value deviates from the expectations and can replace it with the expected value based on the past values. In case of an unknown value it can impute its predicted value.

Figure 3.11 shows an example for the attribute *acc_phone_x*. We see that the application of the filter results in the values being dampened. While this can be useful in some cases, we will not further pursue this avenue for this dataset but stick to our outlier detection with the simple distance based approach and imputation by interpolation. In case a model would be known that relates (multiple) measurements to true values (or states) the filter could certainly be very useful. Also in an online setting (where data comes in on the fly) where we might not have access to a complete dataset, or do not have sufficient time to run our outlier detection algorithms, Kalman filtering could be useful.

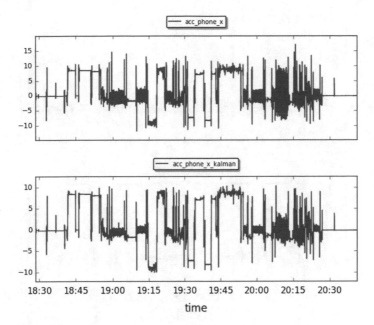

**Fig. 3.11** Kalman filter applied to *acc_phone_x*, the *top graph* shows the original values, the *bottom* the values after applying the Kalman filter (note the differences in scales)

### 3.5.4 Data Transformation

The Butterworth lowpass filter allows us to remove some of the high frequency noise we potentially have in our dataset that might disturb our learning process. We would only expect this for the accelerometers, magnetometers, and the gyroscope. Let us focus on one accelerometer measurement, namely *acc_phone_x*. Given that we know that walking behavior has a frequency between 1 and 1.5 Hz, we look at the influence of filtering data with a frequency above 1.5 Hz. We use an order of 20 in order to guarantee that most of this noise is removed. Figure 3.12 shows the result. We indeed see that the data we obtain after filtering seems much cleaner (and easier to learn from) so we will use the filtered data in the remainder of the book.

Finally, we can try to find the principal components that explain most of the variance in our measurements (note that we consider all measurements/attributes here). If we include the value with respect to those components instead of the raw measurements, it might enable us to achieve better predictive capabilities. Of course, we should only use our predictors and not include our eventual targets. We apply principal component analysis to all our attributes except for the labels (target for classification) and heart rate (target for regression). You can see that we simplify things a bit as we could have also created separate sets for the regression and classification problems, but we do not feel that would add anything given that most of the variance is in the other attributes.

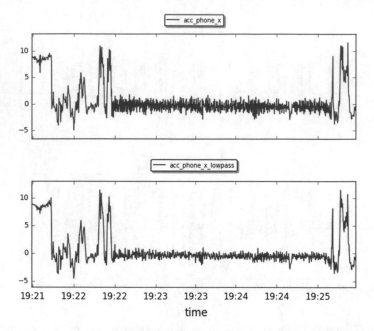

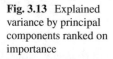

**Fig. 3.12**  Original data (*top*) and filtered data for *acc_phone_x* with frequencies above 1.5 Hz filtered. Note the time scale, we zoomed into part of the data

**Fig. 3.13**  Explained variance by principal components ranked on importance

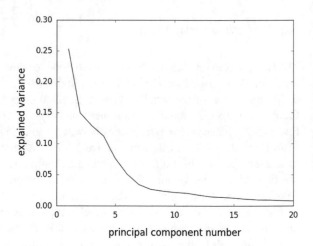

Figure 3.13 shows the explained variance by the different principal components. We clearly observe that the explained variance declines after 7 components (this is sometimes referred to as the elbow). We therefore decide to select 7 components and include the value of each of the components (for each time point) into our dataset.

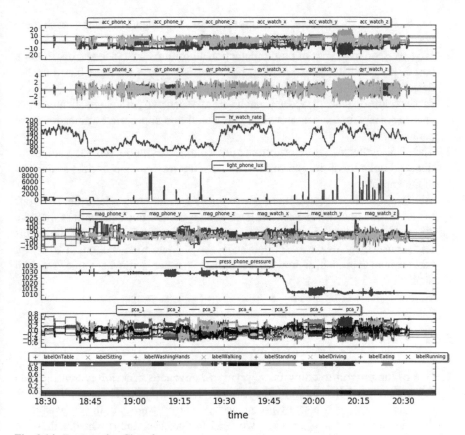

**Fig. 3.14** Dataset after Chap. 3

Finally, the processed dataset after all steps we have just explained is shown in Fig. 3.14. Note the change in the scales of some axes due to the removal of the outliers.

## 3.6 Exercises

### 3.6.1 Pen and Paper

1. In our quantified self setting, we might face datasets of different users. In the description of the techniques in the chapter, we have focused on a single dataset only. When we apply these approaches, should we apply the techniques (and make choices) based on individual user datasets, or should we apply them on the

combination of all datasets? Provide at least two arguments in favor of each these two options.

2. We have seen two types of outlier detection algorithms: distance and distribution based. In what situations would it be better to apply a distance based outlier detection algorithm over a distribution-based approach?

3. In the simple distance-based approach we have seen two parameters, namely $f_{min}$ and $d_{min}$. Explain the way in which you would find appropriate values for these parameters.

4. The local outlier factor algorithm is quite complex. Find out what the computational complexity of the algorithm is and discuss ways to improve the scalability of the approach.

5. We have seen that the Kalman filter assumes a model that relates observations to states. Imagine that we do not have such a model, but that we just map the observations and states one-by-one in a direct way (in fact we have done this for the crowdsignals case). Explain what the Kalman filter will entail when we take such an approach, so how does it update its model and what values would it predict?

## 3.6.2   Coding

1. One of the criteria to allow for applying Chauvenet's criterion is that the data follows a normal distribution. We did not actually verify this in our application of the filter. Study for at least two sequences of sensory values in the crowdsignals data (including the *acc_phone_x* we used in our case study) whether they are indeed normally distributed.

2. To generate Figs. 3.8 and 3.9 we have used the parameter settings described in Sect. 3.5.1. Vary the constant $c$ (smaller and larger values) of the Chauvenet's criterion and study the dependency of the number of detected outliers on $c$. Repeat this for the other three methods presented for outlier detection. Use the source code from book's website, that generated the figures, as a starting point for the analysis.

3. Use a model-based approach to impute the heart rate

4. Similarly to what we have done for our crowdsignals dataset, apply the techniques that have been discussed in this chapter to the dataset you have collected yourself. Write down your observations and argue for certain choices you have made.

5. In line with the previous question, do the same for the case you found covering the data of multiple people. Think about the answer you gave to one of the *pen and paper* questions: how should you tackle the issue with multiple datasets?

# Chapter 4
# Feature Engineering Based on Sensory Data

After having applied the techniques explained in Chap. 3 a cleaned dataset results. While this is a great start, we are not completely ready to apply machine learning algorithms. Especially when we consider the very nature of our problem, we might face data on very different levels. For example, we have accelerometer data but also some Facebook posts. How do we combine this information? We might need to extract some useful features from our dataset to maximize our predictive performance in the end. Of course, we could focus on very specific features (e.g. how can we extract a heart rate from a raw ECG signal), but we are much more interested in generic approaches, that can be applied in various contexts. We will mainly consider features that use the notion of time, both in the time domain and the frequency domain (cf. [78]). In addition, we will discuss features for unstructured data and review the usage of natural language data as this plays a central role in a lot of applications.

## 4.1 Time Domain

Temporal Features in the time domain are extensively used in research focusing on the quantified self [21, 74, 88, 93, 98, 129]. So what is meant by temporal features? Let us start with an example. Imagine that we have a training set in the form of a time series $\mathbf{X}^T$ as shown in Table 4.1. Suppose our target is to perform supervised learning and more precisely to predict whether the person that generates this data (and we all know who we are talking about) is tired or not. If you look closely at the dataset you will see that it will be quite difficult to make this prediction only based on the data at that specific time point. For example, just looking at heart rate or activity is not going to cut it: for instances containing a high heart rate (120) we sometimes observe the target *Tired = yes* and sometimes *Tired = no*. The same holds for the activity level, speed, and activity type. If we consider a history of some time points however, we

© Springer International Publishing AG 2018
M. Hoogendoorn and B. Funk, *Machine Learning for the Quantified Self*,
Cognitive Systems Monographs 35, https://doi.org/10.1007/978-3-319-66308-1_4

**Table 4.1** Example dataset for temporal aggregation

| Time point | Heart rate | Activity level | Speed | Activity type | Tired |
|---|---|---|---|---|---|
| 0 | 45 | Low | 0 | Inactive | No |
| 1 | 120 | High | 10 | Running | No |
| 2 | 45 | Low | 0 | Inactive | No |
| 3 | 120 | High | 10 | Running | No |
| 4 | 120 | High | 9 | Running | Yes |
| 5 | 80 | Medium | 5 | Walking | Yes |
| 6 | 45 | Low | 0 | Inactive | No |
| 7 | 80 | Medium | 5 | Walking | No |

may find informative clues that help us to predict the target: two consecutive heart rates of 80 or above could be a good predictor for our target. Since the vast majority of popular machine learning algorithms will not be able to take advantage of this information as they just consider instances in isolation (note that we will explain learning algorithms that can exploit the temporal dimension in Chap. 8), we need to generate features that encode this temporal information. Engineering features in the time domain is often referred to as the field of temporal data mining (see e.g. [86]). We will discuss approaches for numerical, categorical as well as mixed data.

### 4.1.1   Numerical Data

Let us consider an attribute with numerical values first. Let us say the heart rate, which is the first column in our training data matrix $\mathbf{X}^T$. In general, the values observed over time for attribute $X_i$ can be expressed as $x_1^i, \ldots, x_N^i$ assuming time is discrete and 1 represents the first instance and N the last instance. We now define a new attribute $X_i'$ for which the values represent a summary of the relevant historically observed values. Two terms need to be defined more precisely: the *relevant* historically observed values, and how we *summarize* these observed values. The relevance is defined using a window size $\lambda$ which expresses a number of discrete time points (equal to the number of instances). For the attribute $i$ at time point/instance $t$, $x_t^i$, we determine the new value $x\_new_t^i$ based on the relevant measurements $[x_{t-\lambda}^i, \ldots, x_t^i]$. Note that we can only compute this for $t \geq \lambda$. Thus the window size expresses the number of prior instances or time points that are considered. Of course, the value for $\lambda$ depends on the data and type of measurement and should be set based on domain knowledge or rigorous experimentation. In literature on activity recognition, various window sizes have been reported ranging from instances covering a second of data ([54, 94]), several seconds [13] to 30 s [116]. Gu et al. [54] argue that different windows sizes can be explored, even separate window sizes per feature. Enough about the history:

how do we summarize these selected values? There are again many possibilities: one could consider the mean, median, minimum, maximum, standard deviation, slope, or any other measurement deemed appropriate. Nice overviews of possibilities that have been used in the context of human activity recognition are presented in [78] and [118]. Below, formalizations of commonly used summarization functions are given:

$$x\_mean_t^i = \frac{\sum_{n=t-\lambda}^{t} x_n^i}{\lambda + 1} \tag{4.1}$$

$$x\_max_t^i = max_{t-\lambda \leq n \leq t} x_n^i \tag{4.2}$$

$$x\_min_t^i = min_{t-\lambda \leq n \leq t} x_n^i \tag{4.3}$$

$$x\_std_t^i = \sqrt{\frac{\sum_{n=t-\lambda}^{t} (x\_mean_n^i - x_t^i)^2}{\lambda + 1}} \tag{4.4}$$

Note that using time windows and similar functions to aggregate data should not be completely new to you: we have used it to consolidate our raw data in Chap. 2. Returning to our running example, let us consider the attribute *heart rate* with a window size of $\lambda = 1$. We will use the mean value to summarize the history. The new dataset after adding the attribute is shown in Table 4.2. Note that in practice much more data is needed to draw conclusions on the predictive power of new features.

Of course, we have now just focused on one numerical attribute. Multiple attributes can also be combined if desired. As mentioned earlier, setting appropriate window sizes is an important factor. A way to optimize the window size is proposed in van Breda et al. [28].

**Table 4.2** Example dataset after numerical aggregation

| Time point | Heart rate | Temporal mean heart rate | Tired |
|---|---|---|---|
| 0 | 45 | - | No |
| 1 | 120 | 82.5 | No |
| 2 | 45 | 82.5 | No |
| 3 | 120 | 82.5 | No |
| 4 | 120 | 120 | Yes |
| 5 | 80 | 100 | Yes |
| 6 | 45 | 62.5 | No |
| 7 | 80 | 62.5 | No |

## 4.1.2  Categorical Data

We are now able to exploit the temporal dimension for numerical data, but what about categorical data? If we consider our prior example, the activity type might also be an excellent predictor when considering previous values, e.g. running two times in a row in the last three time points results in *tired = yes*. We cannot just capture this in a numerical value as we could for the numerical data discussed before. First, we need to identify what combinations are useful. Unfortunately there might be a lot of options we can potentially consider as a predictor depending on the number of categories occurring in the attributes. Hence, we need to generate patterns in an intelligent way and need to be selective about which ones we consider. Below we will introduce an example of an algorithm loosely based on (and a simplification of) the algorithm proposed in Batal et al. (cf. [12]) that does precisely what we seek to achieve. We focus on finding temporal patterns in the values of categorical attributes that occur sufficiently frequent. We specify temporal relationships inspired by Allen [6] and focus on values that occur in succession (one before the other, *b*) or occur at the same time point, i.e. co-occur *c*. An example temporal pattern from our dataset is for instance *Activity level = high (c) Activity type = running* or *Activity type = inactive (b) Activity type = running*. The co-occurrence relationship is most valuable if we combine it with the before relationship since learning algorithms will be able to identify the predictive power of the two attributes together, except if we consider co-occurrence relationships before the current time point.

Let us again consider a fixed window size $\lambda$ limiting the history we consider. A notion that will drive our search for these patterns is the *support* of the patterns in our data: how often the pattern occurs in the data compared to the number of time points (or instances) in our dataset. Assume we have found a pattern *pa*. The support is defined as follows:

$$support(pa) = \frac{\sum_{t=t_{start}+\lambda}^{t_{end}} occurs(pa, t - \lambda, t)}{N - \lambda} \tag{4.5}$$

where

$$occurs(pa, t_s, t_e) = \begin{cases} 1 & \textbf{(1)} \; pa \text{ is of the form } X_i = v \text{ and there exists a time} \\ & \text{point between } t_s \text{ and } t_e \text{ where } v \text{ is observed for } X_i, \\ & \textbf{(2)} \; pa \text{ is of the form } pa_1 \text{ (c) } pa_2 \text{ and there exists} \\ & \text{a time point between } t_s \text{ and } t_e \text{ where both } pa_1 \text{ and} \\ & pa_2 \text{ occur, or } \textbf{(3)} \; pa \text{ is of the form } pa_1 \text{ (b) } pa_2 \text{ and} \\ & \text{there exists a time point } t_1 \text{ before } t_2 \text{ both between} \\ & t_s \text{ and } t_e \text{ such that } pa_1 \text{ occurs at } t_1 \text{ and } pa_2 \text{ at } t_2 \\ 0 & \text{otherwise} \end{cases} \tag{4.6}$$

To compute the support, we pass all instances in our data that have enough history, and check whether the patterns occur in the selected history (limited by window size $\lambda$). A pattern is said to occur within the history if it is a simple pattern (i.e. an attribute value combination) for which a time point exists in the given historical time interval where the value is observed. If the pattern is more complex (i.e. contains $(c)$ or $(b)$ constructs), the occurrence of the sub patterns and, additionally, their specified order is verified. Note that we did make a simplification as this definition cannot cope with all nested cases. This is done intentionally to keep the definition simple.

A certain minimal support threshold $\theta$ will be the basis to generate patterns. We simply start to generate all possible patterns of size one first (i.e. attribute value pairs) and check whether their support meets our threshold $\theta$. We then move into a loop where we continuously expand the size of the patterns (referred to as $k$-patterns, where $k$ represents the patterns size). If we would not do this in an efficient way, we would still need to consider a lot of potential patterns. The support, however, comes with a nice property (following the APRIORI algorithm, [5]): we only need to consider $k$-patterns that extend our $(k - 1)$-patterns, and the only way to extend them is by using the $1$-patterns that fulfill our minimum support threshold. This is because the support of a new pattern can never be greater than the support value of the least supported subpattern it includes. This substantially limits our search space. The algorithm is shown in pseudocode below. We refer to the set of generated patterns as $P$:

---

**Algorithm 1:** Temporal Pattern Identification Algorithm

---

$P = \{\}$
$k = 1$
Generate patterns of size 1 (attribute values pairs)
Calculate the support for each pattern and add the ones that reach the threshold $\theta$ to $P$
**while** *True* **do**
    Select the current set of k-patterns $P_k$ from $P$
    Try to extend each element of $P_k$ with an element from $P_1$ using $(c)$ and $(b)$ constructs
    Calculate the support for the new cases
    Add the cases to the set $P$ for which the support $\geq \theta$
    $k = k + 1$
    **if** *no cases have been added* **then**
       | return $P$
    **end**
**end**

---

In the end, the patterns become new attributes (i.e. additions to $X$) for which the value expresses the number of occurrences of the patterns in the relevant history of the instance.

Let us turn to our example to illustrate the whole process. Here, we consider only the attribute *Activity type* as an example. We can see that there are three possible values in our dataset, namely *running*, *walking*, and *inactive*. If we generate 1-*patterns* with a window size of $\lambda = 1$, we find that pattern *Activity type = running* has a support of $\frac{5}{7}$. Note that we ignore the first instance for the calculations when calculating the support, as there is insufficient history available. For computing the support we do not consider how often a pattern occurs within the window for an instance, we just add 1 in case it occurs at least once for an instance and 0 otherwise. We accordingly derive a support of $\frac{5}{7}$ for *Activity type = inactive* and $\frac{3}{7}$ for *walking*. Assuming $\theta = \frac{2}{7}$, we add all to our set $P$. We now move to the 2-*patterns* and try all combinations with all constructs with the 1-*patterns* we have identified. Since we just consider one attribute with a singular value, the co-occurrence is not relevant and the only pattern that meets our threshold is *Activity type = inactive* (*b*) *Activity type = running*. The resulting dataset is shown in Table 4.3. We use a binary count for the field here (i.e. 1 for occurring at least once, 0 otherwise), but could have also used the number of occurrences as value as we mentioned before. As we can see we have generated 4 new features which now can be used for machine learning.

### 4.1.3   Mixed Data

The approaches to add features covering the temporal dimension, that we have introduced above, were very specific for either numerical data or categorical data. Combinations of the two types of attribute values could obviously be beneficial as well. To establish this, we derive categorical values from the numerical data that can in turn be used by applying the above algorithm for categorical data. Batal et al. [12] considers two cases: (1) certain ranges are known that can be used to identify categorical values (think of low, high, and normal blood pressure), or (2) there is only numerical data without an interpretation of what the values mean in the specific context (think of weight, it is difficult to say whether 80 kg is a healthy weight if you do not know how tall a person is). For the former, the translations is obvious. For the latter, the slope can be used as we have previously seen. We did not formally define it, however, we assume we can calculate the slope $x\_slope_t^i$ and say it is *increasing* ($x\_slope_t^i$ is above some threshold), *decreasing* (below the threshold), or *stable*. Although the approach is relatively simplistic, it has been shown to be quite beneficial (see e.g. Kop et al. [75]).

**Table 4.3** Example dataset after adding temporal patterns for attribute Activity type

| Time point | Heart rate | Activity level | Speed | Activity type | Activity type = inactive | Activity type = running | Activity type = walking | Activity type = inactive (b) Activity type = running | Tired |
|---|---|---|---|---|---|---|---|---|---|
| 0 | 45 | Low | 0 | Inactive | - | - | - | - | No |
| 1 | 120 | High | 10 | Running | 1 | 1 | 0 | 1 | No |
| 2 | 45 | Low | 0 | Inactive | 1 | 1 | 0 | 0 | No |
| 3 | 120 | High | 10 | Running | 1 | 1 | 0 | 1 | No |
| 4 | 120 | High | 9 | Running | 0 | 1 | 0 | 0 | Yes |
| 5 | 80 | Medium | 5 | Walking | 0 | 1 | 1 | 0 | Yes |
| 6 | 45 | Low | 0 | Inactive | 1 | 0 | 1 | 0 | No |
| 7 | 80 | Medium | 5 | Walking | 1 | 0 | 1 | 0 | No |

## 4.2  Frequency Domain

Previously, we have identified ways to abstract measurements within a time window to a more aggregated value by means of several aggregation functions (e.g. the mean over time). In this section, we will move to the so-called frequency domain. Remember that we explained the lowpass filter before, which also works in the frequency domain. Let us start with an example to understand this domain a bit better. Imagine we want to recognize whether our friend Arnold is running. While we could use a historic window of the accelerometer measurements and, for instance, the variance we observe there, it might be more natural to look at the periodicity in the measurements of the accelerometer. For example: Do we see that the accelerometer of Arnold shows clear periodic measurements (like a sinusoid) with a frequency similar to a running frequency? For this purpose, we can use Fourier Transformations. We see these types of features in lots of research papers that use sensory data in combination with machine learning (see e.g. [7, 8, 21, 74, 93, 98, 115, 128, 129]).

### 4.2.1  Fourier Transformations

The idea of a Fourier transformation (see e.g. [27]) is that any sequence of measurements we perform can be represented by a combination of sinusoid functions with different frequencies. A typical example includes sound waves, where we have waves at different frequencies that combined make up the sound we hear. The math behind Fourier transformations can be a bit tricky as it involves complex numbers, but we will guide you through it. Let us assume we have a number of sensory measurements for attribute $i$ $[x_{t-\lambda}^i, \ldots, x_t^i]$, where $t$ is our current time point and $\lambda$ the size of the historical window we consider. While there are many variants of Fourier transformations, we will focus on the Discrete Fourier Transform (DFT) as there exists an efficient implementation for this variant. It assumes discrete time (which we also assume throughout this book) and a finite number of measurements (and of course we only consider finite values for our window size $\lambda$). We create sinusoids with different frequencies. Consider $f_0$ as the *base* frequency:

$$f_0 = \frac{2\pi}{\lambda + 1} \tag{4.7}$$

Remember that $\lambda + 1$ is the number of data points in our window (we consider $\lambda$ previous points plus the current time point). We will use multiples of this base frequency, i.e. $k \cdot f_0$, where $k$ is a natural number. The higher the value of $k$ the higher the frequency of the signal. To get from $k$ to a frequency in Hertz we need to know how many datapoints represent a second (called $N_{sec}$):

$$f(k) = k / \left( \frac{\lambda + 1}{N_{sec}} \right) = \frac{k \cdot N_{sec}}{\lambda + 1} \qquad (4.8)$$

$k$ is the number of periods of the sinusoid over our $\lambda + 1$ samples while $\frac{\lambda+1}{N_{sec}}$ is the number of seconds. For each frequency, we also need to specify an amplitude, denoted as $a(k)$ here. We need $\lambda + 1$ different frequencies to represent our original sequence. To be more precise, we need frequencies $\{0 \cdot f_0, \ldots, \lambda \cdot f_0\}$, i.e. $\lambda + 1$ frequencies, starting at 0. Note that in some cases, frequencies are written from $\{-((\lambda + 1)/2)^{-1} \cdot f_0, \ldots, ((\lambda + 1)/2)^{-1} \cdot f_0\}$ but these can be proven to be identical to the frequencies we have mentioned first (we will not bother you with the details). The value of a sinusoid function at frequency $k$ at a time point $t$ is represented as:

$$v_k(t) = a(k) \cdot e^{\mathbf{i} \cdot k \cdot f_0 \cdot t} \qquad (4.9)$$

The $\mathbf{i}$ indicates a complex number (i.e. $\mathbf{i} = \sqrt{-1}$). Do not worry about it if you are unfamiliar with this. What is important to notice is that $e^{\mathbf{i} \cdot k \cdot f_0 \cdot n}$ in fact represents a sinusoid function with the specified frequency $k \cdot f_0$. This follows from Euler's formula:

$$e^{\mathbf{i} \cdot t} = cos(t) + \mathbf{i} \cdot sin(t) \qquad (4.10)$$

We believe everyone loves these kinds of mathematical constructs. We can now compute the value of our measurement at time point $t$ in our window:

$$x_t^i = \sum_{k=0}^{\lambda} a(k) \cdot e^{\mathbf{i} \cdot k \cdot f_0 \cdot t} \qquad (4.11)$$

Are we done? Well, no, although we have fixed the frequencies of the sinusoid functions, we do not yet have the amplitudes $(a(0), \ldots, a(\lambda))$ that match the data in our window. In fact, this a computational problem we should solve. For this purpose, the Fast Fourier Transform algorithm can be used, which runtime is proportional to $(\lambda + 1)log(\lambda + 1))$. We will not go into detail on the working of the algorithm. Of course, we will have different values for the amplitudes over different windows. Therefore we extend the notation for the amplitude a bit with the time window we consider: $a_{t-\lambda}^t(i)$ represents the amplitude of frequency $i$ in the window $[t - \lambda, t]$.

Figure 4.1 shows an example of a series of measurements (we consider a very large window size $\lambda$, the full width of the pattern). We set $N_{sec} = 10$. If we apply our Fourier transformation, we can look at the values of the amplitudes of each of our sinusoid functions with different frequencies (which we can derive from the value for $k$ and our base frequency). Figure 4.2 shows the amplitudes for our case. We see that the sinusoids representing low frequencies are found to be most important in

**Fig. 4.1** Example measurement sequence (time in seconds) Note that the sequence is not displayed as a discretized sequence while this is of course the case

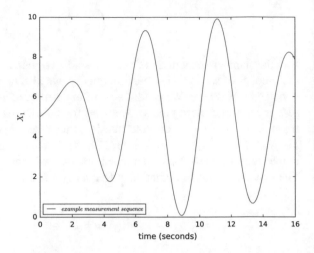

**Fig. 4.2** Amplitudes for example data (*Note* we only selected the part covering frequencies between 0 and 1 Hz as the rest is all zero)

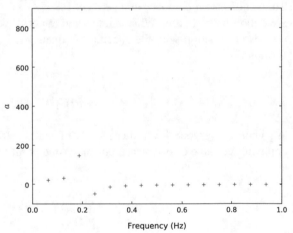

the decomposition, which makes sense given the characteristics of our data (we see a combination of low frequency waves).

As we now know how to transform sequential data into the frequency domain, we will discuss the features that we can distill from it.

### 4.2.2 Features in Frequency Domain

One obvious set of features we can derive in the frequency domain is the amplitude. This amplitude is associated with each of the relevant frequencies that are part of the time window of size $\lambda$ (i.e. $a(0), \ldots, a(\lambda)$). It will have a unique value for each time point we consider:

$$x\_a_t^i(0) = a_{t-\lambda}^t(0)$$

$$\vdots \qquad (4.12)$$

$$x\_a_t^i(\lambda) = a_{t-\lambda}^t(\lambda)$$

Note that we do not refer to the specific attribute $i$ in our notation for the amplitude $a$ to avoid an overly complex notation. Depending on the setting for $\lambda$, this might result in a lot of additional features that are very specific. Features that aggregate these amplitudes and frequencies in a single value have been developed as well. We will explain a few of them here. First of all, we can take the frequency with the highest amplitude. This gives us an indication of the most important frequency in the windows under consideration:

$$x\_max\_f_t^i = f(\underset{k\in[0,\lambda]}{\mathrm{argmax}}\, a_{t-\lambda}^t(k)) \qquad (4.13)$$

A second option is to compute the frequency weighted signal average. The metric provides information on the average frequency observed in the window (given the amplitudes), and might shed a bit more light on the entire spectrum of frequencies contrary to our previous approach that just focused on the most dominant one. It is computed by multiplying the frequencies with their amplitude and normalizing them by the total sum of the amplitudes:

$$x\_f\_weighted_t^i = \frac{\sum_{k=0}^{\lambda} a_{t-\lambda}^t(k) \cdot f(k)}{\sum_{k=0}^{\lambda-1} a_{t-\lambda}^t(k)} \qquad (4.14)$$

Finally, we can derive the power spectral entropy:

$$P_{t-\lambda}^t(k) = \frac{1}{\lambda+1}|a_{t-\lambda}^t(k)|^2 \qquad (4.15)$$

$$p_{t-\lambda}^t(k) = \frac{P_{t-\lambda}^t(k)}{\sum_{i=0}^{\lambda} P_{t-\lambda}^t(i)} \qquad (4.16)$$

$$x\_power\_spec\_entropy_t^i = -\sum_{k=0}^{\lambda} p_{t-\lambda}^t(k) \ln p_{t-\lambda}^t(k) \qquad (4.17)$$

Here, we compute the power spectral density first (squaring the amplitude and normalizing by the number of frequencies), normalize the values to a total sum of 1 such that we can view it as a probability density function, and compute the entropy via the standard entropy calculation. The resulting value represents how much information is contained within the signal. In other words, the power spectral entropy determines, whether there are one or a few discrete frequency standing out of all others.

More features derived from the frequency domain such as the energy of a frequency interval, and the skewedness have been defined and used. See [7] and [118] for an overview.

## 4.3   Features for Unstructured Data

Although the quantified self mainly has to do with data obtained in a structured way (i.e. numerical or categorical attributes), additional information from unstructured data might also be relevant. A lot of unstructured data is collected that can be used in machine learning approaches. Just think of the texts Arnold is sending, the Facebook posts he is generating, and the exercises Bruce is performing to avoid a depression. In this section, we will briefly discuss the extraction of useful attributes from text-based data. This is merely one example of unstructured data. Just think of images and videos, their analysis has recently experienced significant advances through Deep Learning approaches [79].

Here, we will focus on some reasonably simple approaches for natural language processing (NLP) without looking at the semantics of the text. For a more in-depth discussion of NLP approaches the reader is referred to [46]. We will start with the pre-processing of the data, followed by several approaches to form attributes based on different aspects observed in the text.

### 4.3.1   Pre-processing Text Data

Initially, we are faced with raw unstructured text as a value of an attribute $X_i$. Imagine the following fragment:

> Bruce: "I really felt bad yesterday. Got fired at work."

In order to directly create attributes from words or apply some other approaches to extract attributes we initially need to perform a number of basic steps:

1. **Tokenization**: identify sentences and words within sentences.
2. **Lower case**: change the uppercase letters to lowercase.
3. **Stemming**: identify the stem of each word to reduce words to their stem and map all different variations of, for example, verbs to a single term.
4. **Stop word removal**: remove known stop words as they are not likely to be predictive.

Figure 4.3 shows the steps in a graphical form including our example. Note that the stem of a word does not have to be its formal stem as long as there is a single

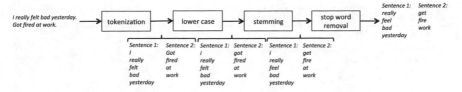

**Fig. 4.3** Simple NLP pipeline

term to map all different variations to. We can perform each of the steps using a variety of NLP tools. We use the following minimalistic notation for text data: $\{x_i^j(1), \ldots, x_i^j(S)\}$ represents the $S$ sentences found in instance $i$ of attribute $j$. Each sentence contains a number of words $W$, denoted within the brackets for the sentence, i.e. $\{x_i^j(1, 1), \ldots, x_i^j(1, W)\}$ represents the words in the first sentence.

### 4.3.2 Bag of Words

After we have identified the words in the various sentences, we are ready to define attributes for the most simple case. We define so-called *n-grams* of words. Here, *n* represents the number of words we consider as a single unit or attribute. A unigram considers single words, a bigram pairs of words, a trigram a combination of three words, etcetera. We look at these combinations in each of our sentences. The approach is called bag of words because we just count the number of occurrences of words irrespective of their order or occurrence (Algorithm 2). If we were to take our example, we would end up with the following unigram attributes: *really, feel, bad, yesterday, get, fire,* and *work.* If we were to take bigrams we would get: *really feel, feel bad, bad yesterday, get fire,* and *fire work.* We refer to each new attribute as $A_j$ and the value of the attribute $j$ for instance $i$ is $a_i^j$ as shown in the algorithm. The value for an attribute is the number of occurrences of the n-gram in the text associated with the instance. Sometimes a binary representation is used. Then we simply replace the count values by binary values to indicate the presence of the n-gram.

### 4.3.3 TF-IDF

An alternative approach is to use the so-called TF-IDF (for Term Frequency Inverse Document Frequency, see [103]) score as a value of an instance $i$ for the n-grams we have identified. This takes into account how unique the n-gram is over the different pieces of text we see in all instances. First, the term frequency (TF) is the number of occurrences of an n-gram in the instance (referred to as a document in this case). We just defined this for the bag of words approach: $a_i^j$. Now, we normalize the value by

---

**Algorithm 2:** Bag of Words (n-grams)

---

$A = \{\}$
$N_{attr} = 1$
**for** $i = 1, \ldots, N$ **do**
$\quad a_i^1, \ldots, a_i^{N_{attr}} = 0$
$\quad$ **for** $s = 1, \ldots, S$ **do**
$\quad\quad$ **for** $w = 1, \ldots, W$ **do**
$\quad\quad\quad$ **if** $w + (n - 1) \leq W$ **then**
$\quad\quad\quad\quad A_{temp} = < x_i^j(s, w), \ldots, x_i^j(s, w + (n - 1)) >$
$\quad\quad\quad\quad$ **if** $A_{temp} \notin A$ **then**
$\quad\quad\quad\quad\quad A = A \cup A_{temp}$
$\quad\quad\quad\quad\quad a_i^{N_{attr}} = 1$
$\quad\quad\quad\quad\quad a_1^{N_{attr}}, \ldots, a_{i-1}^{N_{attr}} = 0$
$\quad\quad\quad\quad\quad N_{attr} = N_{attr} + 1$
$\quad\quad\quad\quad$ **else**
$\quad\quad\quad\quad\quad k = index(A_{temp})$
$\quad\quad\quad\quad\quad a_i^k = a_i^k + 1$
$\quad\quad\quad$ **end**
$\quad\quad$ **end**
$\quad$ **end**
**end**

---

dividing by the number of total instances $N$ that contain the n-gram (this is the IDF part):

$$idf_j = log\left(\frac{N}{|\{i|i \in \{1, \ldots, N\} \wedge a_i^j > 0\}|}\right) \tag{4.18}$$

The higher the number, the more unique the n-gram is (i.e. if it occurs in all documents the value would be the lowest possible: 0). We are now ready to compute the TF-IDF score:

$$tf\_idf_i^j = a_i^j \cdot idf_j \tag{4.19}$$

The score gives more weight to n-grams that are unique compared to the regular counts. This avoids very frequent words to become too dominant in our attributes.

### 4.3.4   Topic Modeling

The raw usage of n-grams results in a fine-granular and large set of attributes. An alternative is to use an algorithm that extracts more high-level topics from the set of texts we have available in our dataset (e.g. the topic "work" or "emotions" for Bruce). Topics are specified by a set of words (all of the words occurring in the text after applying our NLP pipeline: $N_{attr}$) and associated weights $w_i^j$: $topic(k) = \{<A_1, w_k^1 >, \ldots, < A_{N_{attr}}, w_k^{N_{attr}} >\}$. We will give a (hopefully) intuitive explanation how these topics are derived.

To find the topics we can use multiple approaches. We will use Latent Dirichlet Allocation (LDA), cf. [20]. In the approach, we assume a generative process for pieces of text given $k$ topics. This is to say, texts are generated with a certain number of words $W$ (generated by a Poisson distribution) and a distribution over the topics following a so-called Dirichlet distribution (e.g. "Typically, 50% of my writing is work-related, while the other 50% deal with emotions. There aren't any other topics I am writing about."). For each of the $W$ words we select a topic based on the probabilities. Once we have decided about the topic we choose a word according to the multinomial probabality distribution that is specific for this topic (e.g. "job" with a probability of 0.05 for the topic "work").

The key here is to find the probabilities of the words associated with the topics (hence, this is what we started with in the first place; words and weights for topics). We start by assigning a word to a topic at random (giving it a non-zero weight a single topic, and zero in all others) and iteratively improve the weights to maximize the match between our generative process given the topics and associated weights for words and the observed data. The precise details can be found in [20].

Once we have found the weights, we create attributes per topic, and assign a value based on the observed frequencies of words and weights assigned to the words for the topic:

$$topic_k(i) = \sum_{m=1}^{N_{attr}} a_i^m \cdot w_k^m \tag{4.20}$$

We end up with scores for all topics.

## 4.4 Case Study

Let us return to our crowdsignals dataset. As our dataset does not contain any free text, we cannot apply the natural language data approaches. Features in the time and frequency domain might however be very useful.

## 4.4.1   Time Domain

When we explore our dataset in a bit more detail, we do see some relationship between the numerical measurements and the labels or the heart rate. Due to fluctuations in the data and the fact that the changes in the measurements are potentially valuable for prediction tasks, a single measurement at one specific time point might not be enough. Hence, deriving features in the time domain as we have discussed in this chapter could turn out to be very useful. We focus on two ways to aggregate our numerical data in the time domain: We take the standard deviation and the mean over a certain window size. We make this choice based upon our domain knowledge: The standard deviation will say something about the variation in the data (walking results in more variation than standing still for example) while the mean will say more about the general observations over the last time points with more limited influence of a single noisy measurement. Let us focus on the example of the attribute $acc\_phone\_x$ again. A crucial aspect is the selection of the window size. If we select a too small window size, the data might be strongly affected by noise. A too large window size will result in too little variation in our measurements and lack of predictive power. We experiment with different window sizes (20 instances (5 s), 120 instances (30 s), and 1200 instances (5 min), note that window sizes are specified by instances) and explore the influence. Figure 4.4 shows the results. A window size of 20 instances results in a lot of noise, while the 1200 instance window evens out most of our variation. 120 instances is a great middle ground. When considering the figure we can already see the potential in the new values we derived if we just look at the occurrence of labels and values at the different time points. It is obvious to use the aggregation function for similar measurements as well, namely the magnetometer, gyroscope, and the other accelerometer measurements. In addition, we will apply it to the other numerical measurements. For the $press\_phone\_pressure$, and the pca attributes we see similar patterns as we have seen for the accelerometer data. For the $hr\_watch\_rate$ and $light\_phone\_lux$ the mean could filter out some noise. The standard deviation is less obvious but we will see in later chapters whether it is still useful.

Besides numerical attributes we also have one categorical attribute, namely the labels. If we apply our categorical abstraction algorithm with patterns of at most size two, a minimum support of 0.03 (3% of the instances) and a somewhat larger window size of 1200 instances (since we see prolonged periods of the same label a larger window size provides more interesting patterns covering combinations of different labels) we obtain the patterns shown in Table 4.4. The parameter choice is achieved by experimentation: we want to have a good number of patterns, but do not want to get a lot of patterns that hardly occur. While this is not an exact algorithm, experimentation can be beneficial for predicting, e.g. running or walking a number of time points could be predictive for the heart rate. We will therefore include these features in our dataset as well and study the benefit of the newly created columns later on.

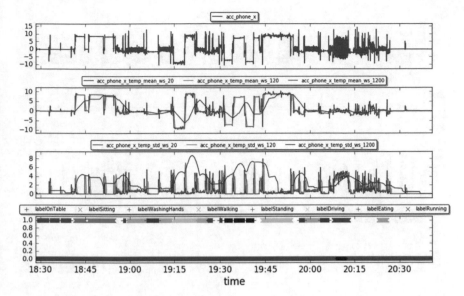

**Fig. 4.4** Numerical temporal aggregation with different window sizes (a window size of 20 resembles 5 s, 120 is 30 s, and 1200 is 5 min)

**Table 4.4** k-patterns found in temporal abstraction

| 1-patterns (7) | 2-patterns (10) |
| --- | --- |
| OnTable, Sitting, Walking, Standing, Driving, Eating, Running | OnTable (b) OnTable, Sitting (b) Sitting, Walking (b) Walking, Walking (b) Standing, Walking (b) Driving, Standing (b) Walking, Standing (b) Standing, Driving (b) Driving, Eating (b) Eating, Running (b) Running |

## 4.4.2 Frequency Domain

Besides the time domain we also explore the frequency domain. We again need to select a window size to compute the amplitudes of the frequencies using a Fourier Transformation. In this case we select a slightly different window size (as 120 instances gives us a huge number of additional features), namely 40 instances, which equals 10 s. This reduces the number of features while the interesting frequencies (e.g. the frequency of walking) are still considered. Figure 4.5 focuses on the attribute *acc_phone_x* and shows the features which aggregate the frequencies that we have discussed before, including the maximum frequency, the frequency signal weighted average, and the power spectral entropy. In terms of the frequency with the highest amplitude, we see that there are activities where the lower frequencies clearly score highest (sitting, driving), while other activities show relatively high frequencies (walking running) while for other activities the picture is less clear. Obviously, it could be more clear cut if we consider the amplitudes of all frequencies. The fre-

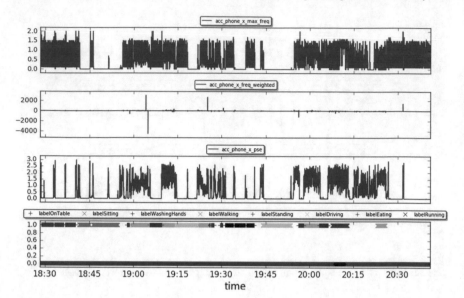

**Fig. 4.5** Frequencies with the aggregated features for the Fourier transformation combined with the labels

quency signal weighted average visually does not provide us with a lot of additional information, we see some extreme outliers causing the other values to be more or less similar in the figure. The power spectral entropy shows a similar pattern as we have seen for the frequency with the highest amplitude, although it seems to contain less noise. For all periodic measurements (all accelerometer, gyroscope, and magnetometer measurements) we add the amplitudes over all frequencies for the 40 instance windows as features and the three aggregate features to the dataset.

### 4.4.3   New Dataset

We have now created quite a few new attributes. However, since we use overlapping time windows the resulting attributes are highly correlated (see Fig. 4.6). Given this overlap, just including all instances might not provide us with new information as only one point in the window differs for adjacent instances. Hence, the aggregated value is likely very similar. This can potentially cause overfitting. Therefore we usually set a maximum overlap for the windows and accordingly remove instances for which this criterion is not met (i.e. we remove intermediate data points). Typically, 50% overlap is allowed for (see e.g. [7, 15]). Due to the limited amount of data we

**Fig. 4.6** Overlapping
windows

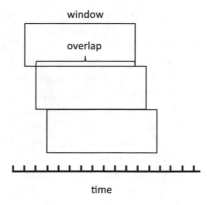

have available, we use 90%. This leaves us with 2895 instances. Of course we do lose some information, but it has been shown to pay off as we have less very similar instances in our set that could cause overfitting.

## 4.5  Exercises

### 4.5.1  Pen and Paper

1. We have seen several functions that summarize numerical values within the time domain to a single number (i.e. mean, standard deviation, minimum, and maximum). Provide an example for all four functions that shows where that specific form of summarization can be useful.
2. Define at least two additional summarization functions for numerical values in the time domain and explain what their added value would be over the four we have already defined in the book. Provide intuitive examples to illustrate your point.
3. The algorithm we have seen for the extraction of temporal patterns from categorical data tries to reduce the complexity by only considering extensions to patterns by attribute value pairs that are sufficiently frequent. Still, the algorithm is quite demanding in terms of computational complexity as we have to figure out the support for a lot of patterns. Think of ways we can use to further reduce this computational complexity.
4. We have discussed a number of features in the frequency domain, including amplitudes for different frequencies, the frequency with the highest amplitude, the frequency weighted signal average, and the power spectral entropy. Find at least two more features we can extract from the frequency domain, and explain their purpose.

5. Sometimes we see negative amplitudes in our Fourier transformations. Explain what these negative values signify.
6. Besides generic features, we might also have dedicated features we engineer for a specific domain. Imagine that we want to learn a model that predicts someones mood based on the amount of social activity. Define three dedicated features that can be useful in this context based on measurements we can potentially be collected from the mobile phone.
7. We have discussed dedicated approaches for handling text based data. One aspect we discussed was to perform stemming on the words to make sure all conjugates of verbs or plural forms of nouns are considered as the same word. Think of one advantage and one disadvantage of using stemming.
8. Imagine that we are working on a supervised learning problem with two classes. One class only occurs in 2% of the cases. When we apply topic modeling, are we guaranteed to get topics that distinguish the different classes well? If not, what could be a solution to solve this problem?

## 4.5.2   Coding

1. Explore the frequency domain features for the crowdsignals dataset in more detail, consider the individual frequencies for the different measurements and see whether you can find interesting patterns. Do you see consistent amplitudes of certain frequencies during the same activities? And how do the amplitudes differ for the different activities?
2. Implement at least two additional metrics in the time domain and the frequency domain in addition to the ones already present in the data (e.g. the ones you have identified in a previous question). Calculate them for the crowdsignals data and discuss their usefulness.
3. Use your own dataset you have collected in Chap. 2 or the dataset you have found on the web and apply the approaches that have been explained in this chapter to identify features. Try different settings for the parameters (e.g. the window size). Report on your findings in a similar way as we have done for the crowdsignals dataset.
4. Find or create a dataset which contains a text component and can be considered a supervised learning problem. Apply the three algorithms that have been explained in this chapter to extract features from the text and explore their relationship to the class value. Would they be useful predictors?

# Part II
# Learning Based on Sensory Data

# Chapter 5
# Clustering

This chapter is devoted to techniques that can provide us with insights in the data, namely whether we can find some structure in terms of clusters in the data. For instance, we might want to identify clusters of locations often visited by Bruce to see what impact a specific location has on his mood. You could also be interested in finding clusters of points that identify different levels of activity for Arnold. Another option is finding clusters of like-minded people, that way we could offer them feedback and support, which seems to work well for their fellow clustermen. We will treat such clustering algorithms in this chapter. The membership of data points or people in a certain cluster might in turn become an attribute for our predictive models later on. We will start by discussing the learning setup.

## 5.1 Learning Setup

We should consider our specific setting of the quantified self before we can start to apply clustering algorithms. You can imagine that there are many people that generate datasets in the quantified self; there might be a lot of "wannabe Arnold's" and a lot of "please do not let me become Bruce" people out there. We will refer to the datasets of $n$ specific people by means of the notation $qs_1, \ldots, qs_n$ resulting in their datasets $\mathbf{X}_{qs_1}, \ldots, \mathbf{X}_{qs_n}$. Furthermore, let $x_{i,qs_j}$ denote the $i$th data point of the $j$th person. When it comes to clustering, we have two levels on which we can cluster: *individual data points* and the level of a *person*. Let us consider an example to make the distinction clear. Assume that Arnold is generating data with his accelerometer. We might be interested to find clusters of data points with respect to the accelerometer data that represent different activities (e.g. a cluster for walking, jogging, cycling, etc.). This would be clustering over individual data points. Of course, you might take the union of data sets of multiple people as a basis for this type of cluster if desired. On the other hand, we might be interested in defining types of people we collect data

© Springer International Publishing AG 2018

M. Hoogendoorn and B. Funk, *Machine Learning for the Quantified Self*,
Cognitive Systems Monographs 35, https://doi.org/10.1007/978-3-319-66308-1_5

from, e.g. a cluster for people who share similar characteristics as Arnold or Bruce. To make this more formal, the points in the clustering space we consider for the two scenarios are:

- *individual data points*:

$$\mathbf{X} = \begin{bmatrix} x_{1,qs_1} \\ \vdots \\ x_{N,qs_1} \\ \vdots \\ x_{1,qs_n} \\ \vdots \\ x_{N,qs_n} \end{bmatrix}$$

- *person*:

$$\mathbf{X} = \begin{bmatrix} \mathbf{X}_{qs_1} \\ \vdots \\ \mathbf{X}_{qs_n} \end{bmatrix}$$

Obviously, this choice has an impact on how to define the distance between the points in the clustering space. We will see this in the next section.

## 5.2   Distance Metrics

Clustering algorithms in general work with a notion of distance between points. We will look at the distances between points for our different setups of our clustering problem.

### 5.2.1   Individual Data Points Distance Metrics

For individual data points there are a lot of commonly used distance metrics. It depends on the nature of the data which one would be appropriate to use. In case we only have numerical data we can use two of the most well known metrics, the *Euclidean* and the *Manhattan* distance. The two distances between data points $x_i$ and $x_j$ are defined as follows:

$$euclidean\_distance(x_i, x_j) = \sqrt{\sum_{k=1}^{p}(x_i^k - x_j^k)^2} \qquad (5.1)$$

$$manhattan\_distance(x_i, x_j) = \sum_{k=1}^{p} |x_i^k - x_j^k| \qquad (5.2)$$

The Euclidean distance corresponds to what we typically just call the distance between two points. The Manhattan distance is an alternative and considers that you can only connect points by moving horizontally or vertically and not diagonally as the Euclidean distance does. It uses a distance function similar to the movement over a grid (like the map of Manhattan, hence the name). An illustration of the difference is shown in Fig. 5.1.

The choice for either one of the two approaches highly depends on the dataset and can mostly only be determined after rigorous experimentation.

A generalization of the above metrics is the so-called Minkowski distance:

$$minkowski\_distance(q, x_i, x_j) = (\sum_{k=1}^{p} |x_i^k - x_j^k|^q)^{\frac{1}{q}} \qquad (5.3)$$

We can see that $minkowski\_distance(1, x_i, x_j) \equiv manhattan\_distance(x_i, x_j)$ and $minkowski\_distance(2, x_i, x_j) \equiv euclidean\_distance(x_i, x_j)$. When considering these distance metrics, one should also consider whether scaling the data is needed or not.

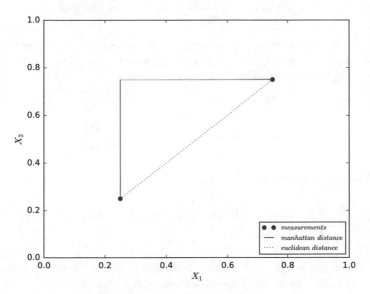

**Fig. 5.1** Difference between Euclidean and Manhattan distance

Otherwise, certain attributes with a high magnitude and spread of observed values might get a very dominant role in the distance calculations.

When we have a combination of numerical and categorical attributes we cannot use the above distance metric. We could transform our categorical attributes into a binary representation, each value of the attribute becoming a new binary attribute. Another metric that can be used is Gower's similarity measure. With Gower's similarity, we only use an attribute $k$ to measure the distance in case it has a value for both instances $i$ and $j$ (i.e. both are not unknown). The similarity for the attribute $k$ is $s(x_i^k, x_j^k)$. It is defined in a different way for different types of variables. The outcome is always scaled to $[0, 1]$. For dichotomous variables (present or not) it is defined as

$$s(x_i^k, x_j^k) = \begin{cases} 1 & \text{when } x_i^k \text{ and } x_j^k \text{ are both present} \\ 0 & otherwise \end{cases} \tag{5.4}$$

For categorical data:

$$s(x_i^k, x_j^k) = \begin{cases} 1 & \text{when } x_i^k = x_j^k \\ 0 & otherwise \end{cases} \tag{5.5}$$

And for numerical data:

$$s(x_i^k, x_j^k) = 1 - \frac{|x_i^k - x_j^k|}{R_k} \text{ where } R_k \text{ is the range of } k \tag{5.6}$$

In order to compare to instances, we compute these functions for all attributes and divide the sum of these values by the number of possible comparisons. This leads us to Gower's similarity measure:

$$gowers\_similarity(x_i, x_j) = \frac{\sum_{k=1}^{p} s(x_i^k, x_j^k)}{\sum_{k=1}^{p} \delta(x_i^k, x_j^k)} \tag{5.7}$$

where $\delta(x_i^k, x_j^k)$ is defined as

$$\delta(x_i^k, x_j^k) = \begin{cases} 1 & \text{when } x_i^k \text{ and } x_i^k \text{ can be compared (i.e. both have values)} \\ 0 & otherwise \end{cases} \tag{5.8}$$

The distance is calculated by taking 1 minus the similarity. This concludes our discussion of distance metrics between individual data points.

## 5.2.2  Person Level Distance Metrics

When we want to cluster datasets of individuals, it becomes a bit more difficult. We now need a distance metric between complete datasets. To determine a suitable metric, we first need to understand how comparable datasets are. If datasets consist of time series that have been measured at the same granularity there are various ways to compute the distance between them. This does not hold for the general case where we do not assume a temporal ordering of the instances. Let us consider the case without temporal ordering first, and then look into comparing time series.

### 5.2.2.1  Non-temporal Distance Metrics

We have a set of variables $X_1, \ldots, X_p$. For each variable $X_i$, we have a number of measurements $N_{qs_j}$ for person $j$: $x^i_{1,qs_j}, \ldots, x^i_{N_{qs_j},qs_j}$. For each person we summarize these values for the variable $X_i$ in a single value, thereby creating a single instance per person. Using these instances we can apply the distance metrics we have explained previously. We can use the mean, standard deviation, minimum, maximum, or whatever summarizing function is deemed appropriate in the domain. For categorical attributes we can create new binary attributes for each value of the original attribute and use a similar approach. While our approach is simple (which is always nice), we might have lost a lot of information in our summarization step.

An alternative is to summarize the measurements for a specific variable of a person by means of its distribution (e.g. for the normal distribution this would be $\mu$ and $\sigma^2$). Hence, we fit a distribution to the data and use the parameter values of the model to summarize the values collected for a person and the difference in these values signifies the distance.

While the previous option is better than just summarizing by means of a single number, a third alternative exists, which is based on statistical tests. Specifically, we can test how the distributions of a variable for two different persons varies for numerical data, resulting in a probability for the assumption that both originate from the same distribution. This probability is often referred to as the $p$-value. The closer the distributions are, the smaller is the $p$-value. So we take $1 - p$ as a distance metric. Any statistical test can be used depending on the distributions. The Kolmogorov-Smirnov test (cf. [73]) could be a nice choice here since it does not make any assumptions about the underlying distribution. A graphical illustration of the three approaches that have been discussed is shown in Fig. 5.2.

### 5.2.2.2  Temporal Distance Metrics

If we want to compare time series that are measured with the same granularity we can use other distance metrics. There are three options as discussed by Liao [81]: *raw data-based*, *feature-based*, and *model-based*.

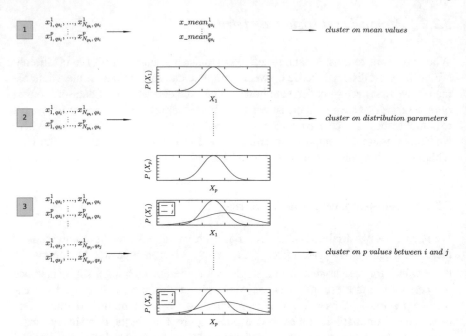

**Fig. 5.2**  Three approaches for calculating non-temporal person-level distances

**Raw-Based Distance Metrics**. Let us consider the *raw data-based* clustering first. For this approach we take the raw series of data for each attribute and define a distance metric. In Fig. 5.3, two series of accelerometer data generated by Arnold and his training buddy Eric are shown. We want to discuss three versions of the raw data-based approach. The first approach focuses on the differences between

**Fig. 5.3**  Two example time series of Arnold and his buddy Eric

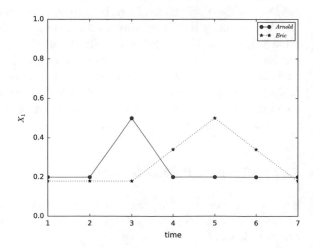

individual data points using for example the Euclidean distance. This is done by taking the vector with all measurements for a specific attribute $l$ over time:

$$euclidean\_distance\_per\_attribute(x^l_{qs_i}, x^l_{qs_j}) = \sqrt{\sum_{k=1}^{N}(x^l_{k,qs_i} - x^l_{k,qs_j})^2} \quad (5.9)$$

Note that this does assume an equal number of points (i.e. we have the same $N$ for both datasets). We can find the overall distance by summing the values over all attributes:

$$euclidean\_distance(x_{qs_i}, x_{qs_j}) = \sum_{l=1}^{p} euclidean\_distance\_per\_attribute(x^l_{qs_i}, x^l_{qs_j}) \quad (5.10)$$

An alternative is to consider the cross correlation between the two different time series. We can use the Pearson correlation coefficient, however the time series we compare might be shifted. We therefore use an approach to handle shifts in the patterns using a *lag*. Consider the two time series in Fig. 5.3 again. We see that they are very similar, only the peak in Eric's values has been shifted in time. Given a shift in time $\tau$, the cross-correlation coefficient is defined as follows:

$$ccc(\tau, x^l_{qs_i}, x^l_{qs_j}) = \sum_{k=1}^{min(N_{qs_i}, N_{qs_j} - \tau)} x^l_{k,qs_i} \cdot x^l_{k+\tau, qs_j} \quad (5.11)$$

We see that the values of person $j$ are shifted and the product of the values is taken. We sum over all available time points for which we can pair them up. The higher the value of this metric, the more the time series are aligned: if peaks align the product will become highest. Finding the value for $\tau$ which maximizes Eq. 5.11 is an optimization problem. In the end, the value for the distance can be defined as:

$$cc\_distance(x_{qs_i}, x_{qs_j}) = \underset{\tau=1,...,min(N_{qs_i}, N_{qs_j})}{argmin} \sum_{k=1}^{p} \frac{1}{ccc(\tau, x^k_{qs_i}, x^k_{qs_j})} \quad (5.12)$$

We optimize the value of $\tau$ over all attributes as the time series should be shifted by the same value across all attributes. A third *raw data-based approach* is called *dynamic time warping (DTW)* (cf. [14]). The cross-correlation coefficient allows for time series that are shifted, but DTW can also take into account that there is a difference in speed between different time series. For instance, if we consider Arnold and Eric their sequences seem to align quite well, except that Eric slowly builds up towards his peak while Arnold does not. We might want to consider these series as relatively close to each other. The DTW algorithm tries to pair measurements (or time points/instances to phrase it differently) of the two time series, i.e. we match each time point in one series to a time point in the other series. These pairs are ordered and can be identified by an index $k$ going from 1 to the time series with the longest length (i.e. $max(N_{qs_i}, N_{qs_j})$). Our pairing does come with the constraints that the time order needs to be preserved (monotonicity condition) and we need to match the first and last points (boundary condition). Let $seq(k, qs_j)$ denote the sequence number of pair $k$ with respect to $qs_j$, then we formalize the monotonicity constraint as follows:

$$\forall l \in 2, \dots, max(N_{qs_i}, N_{qs_j}) : (seq(l, qs_i) \geq seq(l-1, qs_i)) \wedge$$
$$(seq(l, qs_j) \geq seq(l-1, qs_j)) \qquad (5.13)$$

Furthermore, the boundary condition is:

$$seq(1, qs_i) = seq(1, qs_j) = 1 \qquad (5.14)$$
$$seq(max(N_{qs_i}, N_{qs_j}), qs_j) = N_{qs_j} \qquad (5.15)$$
$$seq(max(N_{qs_i}, N_{qs_j}), qs_i) = N_{qs_i} \qquad (5.16)$$

The problem of finding a matching is graphically illustrated in Fig. 5.4. We see the time series of Arnold on the left and the series of Eric at the bottom. Each square in the figure represents a possible pair. Each row is a time point for Arnold while each columns represents a time point for Eric. Moving ahead in time for Arnold is the same as moving up, while moving to the right is moving ahead in time for Eric. To find the pairs, we start at the bottom left and continue by matching points. To find a new pair, we can move up in this figure or move to the right, or both. Given our monotonicity constraint, we can never move down or to the left and our boundary condition requires that we start at the bottom left and end at the top right. The blue squares give an example of a path. The procedure to determine the minimum cost per position in the figure is shown below. In the end, the minimum cost in the upper right corner is returned as the dynamic time warping distance.

**Fig. 5.4** Example dynamic time warping for example series of Arnold and Eric

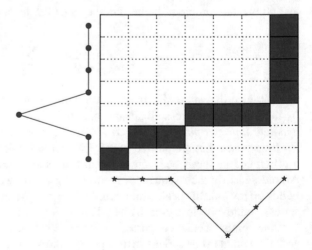

---

**Algorithm 3:** Dynamic Time Warping

dtw_distance($\mathbf{X}^{\mathcal{T}}_{\mathbf{qs}_i}, \mathbf{X}^{\mathcal{T}}_{\mathbf{qs}_j}$) :

**for** $k \in 1, \ldots, N_{qs_i}$ **do**
| cheapest_path($k, 0$) = $\infty$
**end**
**for** $k \in 1, \ldots, N_{qs_j}$ **do**
| cheapest_path($0, k$) = $\infty$
**end**
cheapest_path($0,0$) = 0
**for** $k = 1, \ldots, N_{qs_i}$ **do**
| **for** $l = 1, \ldots, N_{qs_j}$ **do**
| | $d = $ distance($x_{k,qs_i}, x_{l,qs_j}$)
| | cheapest_path = $d + min(\{$cheapest_path($k - 1, l$), cheapest_path($k, l - 1$), cheapest_path($k - 1, l - 1$)$\}$)
| **end**
**end**
**return** *cheapest_path($N_{qs_i}, N_{qs_i}$)*

---

The algorithm states that we cannot move outside of our time series (giving an infinite value for moving before the first point). We then calculate the minimum cost for each position in our search space (i.e. the squares shown in Fig. 5.4). This is the cost of the difference between the values in that square and the cheapest path that leads towards it. We typically use the Euclidean distance as the distance metric. Many improvements have been applied to this algorithm, for instance, limiting the maximum difference in time points of pairs, and more efficient calculations (e.g. the Keogh bound [70]). With this approach, we do not consider individual attributes but focus on the distance between the values of all attributes since we are trying to match time points over the whole dataset. According to some domain knowledge, it would also be possible to consider attributes on an individual basis and perform DTW individually if desired. Although we have not discussed categorical attributes in the context of raw data-based approach, they can be treated in the same way as we have mentioned before, by creating binary attributes per category.

**Feature-Based Distance Metrics**. As said before, we can also take a *feature-based* approach to comparing two time series. In order to do so, we extract features from the time series. For this purpose we can employ the same distance metrics as we have explained for the general case **X** by simply ignoring the temporal ordering. Alternatively, features similar to Chap. 4 can be used. To compare two time series we finally compare the derived features.

**Model-Based Distance Metrics.** For the *model-based* approach we fit a model our time series (e.g. a time series model as explained later in Chap. 8) and use the parameters of the model for the characterization of the time series. We compute the difference between these parameters and use it as the distance between persons.

## 5.3   Non-hierarchical Clustering

We have defined a number of distance metrics for both the cases of individual points
and person level datasets. Given these distance measures we can now start clustering.
For convenience we will use the notation for the individual data points.

We will start with the algorithm called *k-means clustering* [82]. In this approach,
a predefined number of clusters $k$ is found. Each cluster can be identified by a cluster
center. The cluster centers are initially set randomly and then refined in a loop. The
algorithm is shown in Algorithm 4.

---

**Algorithm 4:** k-means clustering

---
**for** $i = 1, \ldots, k$ **do**
  |   centers[k] = random point in the clustering space
**end**
prev_centers = []
**while** *prev_centers != centers* **do**
  |   prev_centers = centers
  |   cluster_assignment = []
  |   **for** $i = 1, \ldots, N$ **do**
  |    |   cluster_assignment[i] = $\text{argmin}_{j=1,\ldots,k} \text{distance}(x_i, \text{centers}[j])$
  |   **end**
  |   **for** $j = 1, \ldots, k$ **do**
  |    |   centers[j] = $\frac{\sum_{l \in \{i | \text{cluster\_assignment}[i]=j\}} l}{|\{i | \text{cluster\_assignment}[i]=j\}|}$
  |   **end**
**end**

---

In the main algorithm loop we compute which cluster each point belongs to. This
cluster is selected based on the minimum distance between the point and the cluster
center. Once we have calculated this for all data points, we recompute the center of
the cluster by taking the average over all data points in the cluster. This continues
until the centers do not change anymore (or only change with a very small value).
This approach is quite intuitive for individual data points and approaches where we
aggregate datasets to single points.

Let us consider an example from the quantified self domain. Imagine we have
data points that represent two dimensions of the accelerometer data and we want to
cluster them in two clusters (i.e. $k = 2$). Figure 5.5 shows an example of the first steps
of the k-means clustering algorithm. Here, the blue and red points are the data points
while the black points are the cluster centers. After step 4, only one more update of
the centers is required before the centers stabilize and the algorithm terminates. For
distance metrics that are computed between whole datasets (i.e. the person level) and
are raw-based k-means clustering is neither appropriate nor intuitive (what does the
center of a set of datasets mean?). It has also been shown that k-means clustering
does not always work well with dynamic time warping (cf. [90]).

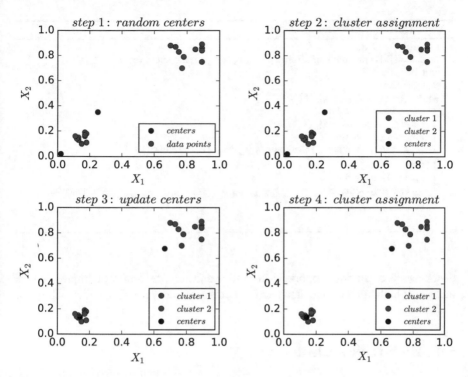

**Fig. 5.5** Example of first steps of k-means clustering algorithm (with $k = 2$)

An alternative approach is k-medoids (cf. [67]). Instead of assigning cluster centers that are averages over the points belonging to the cluster, the k-medoids algorithm selects points from the dataset as cluster centers. This solves our previous problem we have identified with the person level data and also works well with dynamic time warping. The algorithm is very similar to the k-means clustering and is shown in Algorithm 5. The difference appears where the cluster center is assigned. Here, the actual point that minimizes the distance to all points in the cluster is selected.

Let us briefly take a step back: Without explicitly referring to clustering we have already seen another approach in Sect. 3.1.1.2 that supports clustering, that is Gaussian mixture models. In Sect. 3.1.1.2, we used this approach to identify outliers. In order to do so, we assumed that the data points were generated from a probability distribution that was constituted by a number of independent normal distributions and identify outliers with respect to this probability distribution. Implicitly we assumed that each data points, that is not an outlier, belongs to one of the independent normal distributions. Thus, each of these independent normal distributions can be interpreted as a cluster. What is the difference between mixture models and k-means/k-medoids clustering? For k-means/k-medoids we did not assume a probability distribution that generated the data points. Both approaches are appropriate for different applications: While mixture models are used for outlier detection or estimating probability densi-

---

**Algorithm 5:** k-medoids clustering

---

**for** $i = 1, \ldots, k$ **do**
 |    centers[k] = random point from $x_1, \ldots, x_N$ not part of centers yet
**end**
prev_centers = []
**while** *prev_centers != centers* **do**
 |    prev_centers = centers
 |    cluster_assignment = []
 |    **for** $i = 1, \ldots, N$ **do**
 |      |    cluster_assignment[i] = $\text{argmin}_{j=1,\ldots,k}$distance($x_i$, centers[j])
 |    **end**
 |    **for** $j = 1, \ldots, k$ **do**
 |      |    centers[j] = $\text{argmin}_{l \in \{i | \text{cluster\_assignment}[i]=j\}} \sum_{q \in \{i | \text{cluster\_assignment}[i]=j\}}$ distance($x_l, x_q$)
 |    **end**
**end**

---

ties, k-means clustering supports visualization and compression. By compression we mean, that high dimensional data points are represented by the cluster they belong to.

## 5.4  Hierarchical Clustering

The approaches we have seen so far require a predefined number of clusters and we only considered cluster that did not overlap. In hierarchical clustering, however, we drop both requirements. We either start with one big cluster and sequentially refine it (*divisive clustering*) or start with each instance in its own cluster and combine clusters. The latter is called *agglomerative clustering*. Before we dive into the details of the approach, let us consider the end-product of such a clustering which is often a *dendrogram*. The dendrogram shows the results of clustering on different levels, going from one cluster at the top to the most refined clusters at the bottom. This allows us to select the level of clustering which is appropriate for the domain. For an extensive overview of the approaches, see [68].

Figure 5.6 shows an example based on datasets of a number of quantified selves. Each depth of the dendrogram represents a cutoff value for our algorithms. Let us look into both approaches in a bit more detail.

### 5.4.1  Agglomerative Clustering

In agglomerative clustering, we start with each data point being an independent cluster and combine them until all data points constitute one cluster. Which clusters to combine depends on the distance between clusters. We assume to have a distance function between two data points $x_i$ and $x_j$ (which we have discussed extensively

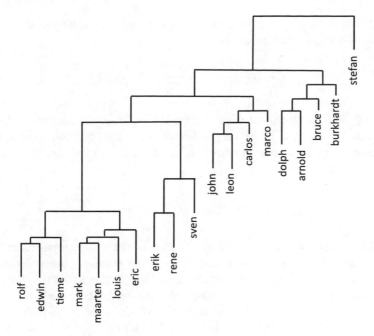

**Fig. 5.6** Example of a dendrogram

before) noted as *distance*$(x_i, x_j)$. Given that we have clusters $C_k$ and $C_l$ we can define the distance between clusters in a variety of ways. We will discuss three here. First, the *single linkage* defines cluster distance as:

$$d_{SL}(C_k, C_l) = \min_{x_i \in C_k, x_j \in C_l} distance(x_i, x_j) \qquad (5.17)$$

Hence, we take the distance between the two closest points as the distance between the clusters. The opposite is called *complete linkage* and takes the distance between two points that are farthest apart:

$$d_{CL}(C_k, C_l) = \max_{x_i \in C_k, x_j \in C_l} distance(x_i, x_j) \qquad (5.18)$$

Third (and naturally) we can use the average distance between the points in the clusters referred to as the *group average*:

$$d_{GA}(C_k, C_l) = \frac{\sum_{x_i \in C_k} \sum_{x_j \in C_l} distance(x_i, x_j)}{|C_k| \cdot |C_l|} \tag{5.19}$$

A slightly different criterion is Ward's method [124]. It defines the distance between clusters as the increase in the standard deviation when the clusters are merged. Assume that $m_{C_i}$ represents the center of cluster $C_i$, then the distance is defined as:

$$d_{Ward}(C_k, C_l) = \sum_{x_i \in C_k \cup C_l} ||x_i - m_{C_k \cup C_l}||^2 - \sum_{x_j \in C_k} ||x_j - m_{C_k}||^2 - \sum_{x_n \in C_l} ||x_n - m_{C_l}||^2 \tag{5.20}$$

Each one of the between cluster distance metrics comes with its own pros and cons, but in general the latter two provide a nice middle ground between a very strict condition for joining clusters (complete linkage) and a very loose one (single linkage). The algorithm to merge clusters is expressed in Algorithm 6. It works by means of a predefined threshold *th* for the maximum distance in order to still merge clusters. Intuitively, this threshold represents a horizontal position in the dendrogram, resulting in a certain division in clusters.

---

**Algorithm 6:** Agglomerative clustering

---

clusters = { }
**for** $x_i \in \mathbf{X}$ **do**
  | clusters = clusters + $\{x_i\}$
**end**
**while** *True* **do**
  | $C_k, C_l = \text{argmin}_{C_k \in \text{clusters}, C_l \neq C_k \in \text{clusters}} \, d(C_k, C_l)$
  | **if** $d(C_k, C_l) > th$ **then**
  |   | **return** *clusters*
  | **else**
  |   | clusters = clusters \ $\{C_k\}$
  |   | clusters = clusters \ $\{C_l\}$
  |   | clusters = clusters + $\{C_k + C_l\}$
**end**

---

We can see that we start with clusters of one data point and merge clusters until we can no longer find clusters that are sufficiently close.

## 5.4.2  Divisive Clustering

As said, divisive clustering works right in the opposite direction of agglomerative clustering. Therefore we start with a single cluster. Let us define the dissimilarity of a point to other points in its cluster:

$$dissimilarity(x_i, C) = \frac{\sum_{x_j \neq x_i \in C} distance(x_i, x_j)}{|C|} \tag{5.21}$$

Using this metric we can split a cluster $C$ by considering the point with the greatest dissimilarity to the cluster. We create a new cluster $C'$ for this and continue moving the most dissimilar points from cluster $C$ to $C'$. We stop when there is no point left that is less dissimilar to the points in cluster $C'$ than it is to the remaining points in cluster $C$. So what cluster should we select for the process we have just described? Various criteria have been defined, one being the cluster with the largest diameter (cf. [68]):

$$diameter(C) = \max_{x_i, x_j \in C} distance(x_i, x_j) \tag{5.22}$$

In other words, the diameter is the maximum distance between points in the cluster. This is used in Algorithm 7.

---

**Algorithm 7:** Divisive clustering

---

clusters = $\{\{x_1, \ldots, x_N\}\}$
**while** $|clusters| < N$ **do**
    $C = \text{argmax}_{C \in clusters} diameter(C)$
    clusters = clusters $\setminus C$
    most\_dissimilar\_point = $\text{argmax}_{x \in C}$ dissimilarity$(x, C)$
    $C = C \setminus$ most\_dissimilar\_point
    $C_{new} = \{$most\_dissimilar\_point$\}$
    point\_improvement = $\infty$
    **while** *point\_improvement > 0* **do**
        $x = \text{argmax}_{x \in C}$ dissimilarity$(x, C)$
        distance\_C = dissimilarity$(x, C)$
        distance\_$C_{new}$ = dissimilarity$(x, C_{new})$
        point\_improvement = distance\_$C_{new}$ - distance\_C
        **if** *point\_improvement > 0* **then**
            $C = C \setminus x$
            $C_{new} = C_{new} + x$
        **end**
    **end**
    clusters = clusters + $C + C_{new}$
**end**
**return** *clusters*

---

Although we do not explicitly have the threshold set in this algorithm but work on a cluster by cluster basis, a dendrogram as we have seen before is the result of this procedure.

## 5.5  Subspace Clustering

While the approaches above are simple, intuitive, and in general work quite nicely, they do have some disadvantages when it comes to our quantified self setting. We might have a huge attribute space (we measure more and more around ourselves) and this causes several problems: (1) our approaches will take a long time to compute, (2) calculating distances over a large number of attributes can be problematic and distances might not distinguish cases very clearly, and (3) the results will not be very insightful due to the high dimensionality. Hence, we need to define a subset of the attributes (or subspace) to perform clustering. We could do this manually or use dimensionality reduction approaches such as Principal Component Analysis. However, the manual approach would need a lot of experimentation, while the latter would not provide intuitive results, as we transform the attributes that initially had a meaning to a less meaningful new space. Methods from *subspace clustering* come to rescue here. We will explain one specific subspace clustering algorithm, namely the *CLIQUE* algorithm [4].

Starting point for our explanation is that we create so-called *units* that partition the data space. Hereto, we split the range of each variable up into $\epsilon$ distinct intervals. This is exemplified for the dataset shown in Fig. 5.7. The splits are shown by the dotted lines.

Assuming $k \leq p$ dimensions or attributes (we can take subsets of the attributes), a unit $u$ is defined by means of boundaries per dimension: $u = \{u_1, \ldots, u_k\}$.

**Fig. 5.7** Example of units in subspace clustering

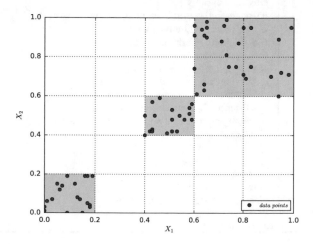

A bound is provided for each dimension for the unit. $u_i(l)$ is the lower bound and $u_i(h)$ the upper bound for dimension $i$. A point belongs to a unit when it falls within all bounds:

$$belongs\_to(x, u) = \begin{cases} 1 & \forall_{i \in 1,...,k} u_i(l) \le x^i < u_i(h) \\ 0 & \text{otherwise} \end{cases} \tag{5.23}$$

For each unit we can define the *selectivity* and *density* of the unit $u$:

$$selectivity(u) = \frac{\sum_{i=1}^{N} belongs\_to(x_i, u)}{N} \tag{5.24}$$

$$dense(u) = \begin{cases} 1 & selectivity(u) > \tau \\ 0 & \text{otherwise} \end{cases} \tag{5.25}$$

In other words, the selectivity is defined as the fraction of instances belonging to the unit $u$ out of all instances, while a unit is called dense when the selectivity exceeds some threshold $\tau$. In Fig. 5.7, we choose $\tau$ so that at least one instance is in the unit. This results in 6 dense units. These units are depicted in grey. As said, the units do not have to cover all dimensions (in fact we would rather like it if they do not). Assuming a unit covers $k$ dimensions (with $k \le p$) we define a cluster as a maximal set of connected dense units. Units $u_1 = \{r_1, \ldots, r_k\}$ and $u_2 = \{r'_1, \ldots, r'_k\}$ are directly connected when they have what is called a common face (i.e. when they share a border of a range on one dimension and have the same ranges on the others):

$$common\_face(u_1, u_2) = \begin{cases} 1 & \exists i \in 1, \ldots, k : (r_i(l) = r'_i(h) \vee r_i(h) = r'_i(l)) \wedge \\ & \forall j \ne i \in 1, \ldots, k : (r_i = r'_i) \\ 0 & \text{otherwise} \end{cases}$$
$$\tag{5.26}$$

Furthermore, indirect connections are also considered. Overall units are connected in a cluster when:

$$connected(u_1, u_2) = \begin{cases} 1 & common\_face(u_1, u_2) \vee \\ & \exists u_3 : common\_face(u_1, u_3) \wedge common\_face(u_2, u_3) \\ 0 & \text{otherwise} \end{cases}$$
$$\tag{5.27}$$

If we consider dense units as defined previously that are connected in our figure we obtain three clusters. These clusters can be specified by means of ranges of values that make up the region. For example, the upper right cluster can be expressed as $(0.6 \le X_1 < 1) \wedge (0.6 \le X_2 < 1)$ but also by $(0.6 \le X_1 < 0.8) \wedge (0.8 \le X_1 < 1) \wedge (0.6 \le X_2 < 1)$. We call the *minimal description* of a cluster the smallest set of regions that still covers all units in the cluster. In our case this would be the first description. Alright, the stage is set. How do we find units that are dense over all these different dimensions? The first problem we tackle is finding these units efficiently. Given the size of our search space, we cannot make calculations for each possible unit in each possible subset of our dimensions. To reduce the search space, we use the fact that a unit $u$ can only be dense in $k$ dimensions if all units of $k-1$ dimensions that are a subset of the constraints for unit $u$ are also dense. This makes sense as the unit $u$ covers a smaller part of the data space (it splits the data up in an additional dimension) and thus can never have more data instances in it. Thus we start with 1 dimension and work our way up to more dimensions. The generation of candidate units with $k$ dimensions given the units with $k-1$ attributes known to be dense is expressed in Algorithm 8. Here, $C_k$ denotes the set of candidate units of dimension $k$ and $R_{k-1}$ stands for the set of dense units of dimensions $k-1$. $u_i(a)$ refers to the name of the $i$th attribute of unit $u$. We look for two units that are part of the dense units in dimensions $k-1$, of which the attributes and bounds overlap in $k-2$ units and add the two non overlapping attributes (and associated ranges) to create a unit with dimension $k$, so we essentially create a unit with one additional attribute compared to the two units we have identified for dimension $k-1$. An ordering is assumed (indicated by $<$) to avoid doubles. For all candidates generated based on this algorithm, we compute whether they are dense or not.

---

**Algorithm 8:** Dense unit candidate generation from $k-1$ to $k$ attributes

$C_k = []$
**for** $u, u' \in R_{k-1}$ **do**
   **if** $u_1(a) == u'_1(a) \wedge u_1(h) == u'_1(h) \wedge u_1(l) == u'_1(l) \wedge \cdots \wedge$
   $u_{k-2}(a) == u'_{k-2}(a) \wedge u_{k-2}(h) == u'_{k-2}(h) \wedge$
   $u_{k-2}(l) == u'_{k-2}(l) \wedge u_{k-1}(a) < u'_{k-1}(a)$ **then**
      $C_k = C_k + < u_1, \ldots, u_{k-1}, u'_{k-1} >$
   **end**
**end**
**return** $C_k$

---

While this already helps in terms of computation, we can further improve matters by focusing on subspaces (i.e. subsets of the attributes) that contain a significant proportion of the overall data points. Assuming subspaces $S_1, \ldots, S_n$ we compute the number of points that belong to the units that are part of the subspace and that are part of a dense unit (following our previous algorithm). We call this the *coverage*:

$$coverage(S_i) = \sum_{u \in S_i} \left( dense(u) \cdot \sum_{i=1}^{N} belongs\_to(x_i, u) \right) \tag{5.28}$$

Only subspaces with a large coverage will be selected. We therefore sort the subspaces according to their coverage score: $S_1, \ldots, S_n$ where $S_1$ has the highest coverage. We want to create a set of selected subspaces $I$ and those we want to prune $P$. For this purpose we select a point $i$ in the ordered list at which we split it: $I = \{S_1, \ldots, S_i\}$ and $P = \{S_{i+1}, \ldots, S_n\}$. We choose $i$ based on a heuristic that aims at minimizing the number of bits required to send information on the values of the coverage for all subspaces. The lower the amount of information we need to send, the better the split we have obtained. Given split $i$ the minimum information we are required to send is the average coverage of the subspaces in $I$ (called $\mu_I(i)$) and $P$ ($\mu_P(i)$) and the deviation of the coverage of each subspace from this average. Since we send the information in bits we take the $\log_2$ of these values:

$$\mu_I(i) = \frac{\sum_{j=1}^{i} coverage(S_j)}{i} \tag{5.29}$$

$$\mu_P(i) = \frac{\sum_{j=i+1}^{n} coverage(S_j)}{(n-i)} \tag{5.30}$$

$$information(i) = \log_2(\mu_I(i)) + \log_2(\sum_{j=1}^{i} |coverage(S_j) - \mu_I(i)|) +$$

$$\log_2(\mu_P(i)) + \log_2(\sum_{j=i+1}^{n} |coverage(S_j) - \mu_P(i)|) \tag{5.31}$$

All we need to do is find the value of $i$ that minimizes this sum. The idea behind this approach is that the pruned set will contain all subspaces with very low coverages (and thus low variation and information).

We now have a selection of subspaces of certain dimensions and we know how to effectively compute dense units within those subspaces. A logical next step is to find units that when combined make up clusters in a certain subspace. We can do this using a depth first search like algorithm. Part of the procedure is shown in Algorithm 9. We start with a cluster number $n$ and a unit $u$. We then go through the $k$ dimensions of the current subspace and look for neighbors on the left and right of the point to see whether they are dense as well and not part of a cluster yet. If this is the case, we continue along that avenue. This results in the assignment of units to cluster $n$. Once we have found the clusters, we describe them in terms of their ranges of values (as we previously indicated) and find a minimal way to do so. Agrawal et al. [4] discuss in detail how this can be done in an efficient way.

After all these steps have been performed we end up with a selection of suitable subspaces and a description of the clusters in those subspaces.

## 5.6 Datastream Clustering

Although subspace clustering helps us solve the dimensionality issue, some challenges and restrictions remain. All of the algorithms we have seen so far make some strong assumptions [1]:

---

**Algorithm 9:** Cluster generation

---

find_neighbors(**u**, **n**) :
cluster($u$) = $n$
**for** $j = 1, \ldots, k$ **do**
    $u^l = u$
    $u_j^l(h) = u_j(l)$
    $u_j^l(l) = u_j(l) - 1$
    **if** $dense(u^l) \wedge cluster(u^l) = unknown$ **then**
       |  find_neighbors($u^l$, $n$)
    **end**
    $u^r = u$
    $u_j^l(l) = u_j(h)$
    $u_j^l(h) = u_j(h) + 1$
    **if** $dense(u^h) \wedge cluster(u^h) = unknown$ **then**
       |  find_neighbors($u^l$, $n$)
    **end**
**end**

---

- They assume that an unlimited amount of data can be stored, such that multiple passes can be performed, e.g. in the k-means clustering. In our setting of the quantified self, storing all sensory data in a highly fine-grained manner might exceed the storing capacity of our mobile devices. A central repository would be an option, however, imagine Arnold's accelerometer data with more than 100 samples per second. Uploading this data would take way too much bandwidth.
- Another assumptions is that all data should be treated in the same way. However, the underlying mechanisms generating the data might evolve over time, requiring models to become outdated. For instance, Bruce might improve his ability to control his blood glucose level. This is referred to as *temporal locality* or *concept drift*. Note that this is different from our temporal/non-temporal predictors we have seen in the previous section.

Within the domain of *data stream mining* algorithms are being developed that no longer build on the assumptions above. We will give a few example for clustering approaches which tackle some of the problems listed earlier. One approach is to maintain a window of a particular size $n$ and cluster only on the last $n$ elements. Each new arriving instance then replaces an element in our window (e.g. the oldest) or we replace elements only with a certain probability less than 1 to avoid having to run the algorithms repeatedly.

An alternative approach is to store cluster centers for chunks of data and continue abstracting over these centers when our dataset grows [55]. This might sound a bit vague, but it is actually pretty straightforward. Imagine that we select the first $m$ elements in our data. We cluster these $m$ elements by means of k-medoids and identify $k$ instances as centers. We assign a weight $w$ to each center based on the number of instances that are part of the cluster. We continue this process with the next chunk of data of size $m$ until we have done this $\frac{m}{k}$ times. We now have a new dataset of $m$ centers and their associated weights. We cluster these centers into $k$ clusters again based on the weights of the medoids we had previously assigned. We can continue this process for multiple levels, and we only store the medoids, weights for each level. Hence, we greatly reduce our need for storing lots of data.

Another alternative is to use the mixture of normal models that we have discussed earlier. For this purpose, you can imagine that part of the data, for example the last $n$ data points, is used to estimate an initial mixture of normals. This results in a compressed representation of the $n$ data points in terms of the means and standard deviations of the mixture components. As new data comes in, the density estimates are updated. It has been shown that this is a very efficient way to handle data stream clustering [109].

## 5.7 Performance Evaluation

Clustering is not as clear cut in terms of performance metrics compared to supervised learning approaches which we will discuss in the next three chapters. So how do we know that we have found a good clustering? Well, there are some metrics that can help out. We will discuss one of the most prominent examples here, namely the *silhouette score* [101]. This cannot only help to evaluate the clustering but also to find good values for the parameters of various clustering approaches (e.g. $k$ for k-means and k-medoids). We start by defining the average distance of a point to the other points in its cluster:

$$a(x_i) = \frac{\sum_{\forall x_j \in C_l} distance(x_i, x_j)}{|C_l|} \text{ where } x_i \in C_l \qquad (5.32)$$

In addition, the silhouette score uses the average distance to the points in the cluster closest by

$$b(x_i) = \min_{\forall C_m \neq C_l} \frac{\sum_{\forall x_j \in C_m} distance(x_i, x_j)}{|C_m|} \text{ where } x_i \in C_l \qquad (5.33)$$

We can now define the silhouette:

$$silhouette = \frac{\sum_{i=1}^{N} \frac{b(x_i) - a(x_i)}{max(a(x_i), b(x_i))}}{N} \qquad (5.34)$$

It compares the distances $a(x_i)$ and $b(x_i)$ and divides it by the maximum of the two. Hence, it provides a measure on how tight the clusters themselves are relative to the distance to the clusters closest to them. The score can range from $-1$ to $1$, where $-1$ is clearly the worst score one could obtain as apparently the distances from points to other clusters are lower than those between points within the cluster. The closer to 1 the better it is, since close to 1 represents a low value for the $a(x_i)$'s (tight clustering) and high values for $b(x_i)$'s (cluster are far apart).

## 5.8 Case Study

Let us go back to our crowdsignals dataset. While our goal is not explicitly set to finding interesting clusters, clustering could still contribute to solving our problem (predict the label or the heart rate): if we can cluster our measurements in such a way that the membership of a cluster is predictive for the target, it would be a great contribution, making the presence of an instance in a cluster a new attribute. In addition, it provides us with insights into our data that will help to make choices in the next chapters. We will try several of the clustering approaches we have discussed in this chapter. While we have ample options to choose from in terms of distance metrics and a learning setup, we take a rather straightforward one. We will use the Euclidean distance as a distance metric (it is easy to understand, applicable to both k-means and k-medoids, and fast). For the selection of the learning setup we do not have much choice: we just have the dataset of a single quantified self. Therefore we aim to cluster instances in our data instead of selecting the person level. For the clustering we will focus on the accelerometer of the phone, i.e. the attributes *acc_phone_x*, *acc_phone_y*, and *acc_phone_z*. Our task is to find clusters that might be indicative for the type of activity being conducted (though we do not use the label information to generate the clusters). Of course, this selection of measurements is a bit arbitrary, but it is a set of measurements that is representative for most of our measurements.

### 5.8.1 Non-hierarchical Clustering

First, let us dive into non-hierarchical approaches, namely k-means and k-medoids.

#### 5.8.1.1 K-Means

Given our dataset, we need to figure out the best setting for the number of clusters ($k$) first. For this, we run the algorithm with different values for $k$ (ranging from 2 to 9 clusters) and measure the silhouette to judge the quality of the clustering. The result is shown in Fig. 5.8. Note that we only do some exploratory runs with a limited number of random initializations per setting to get an idea on the best value for $k$. From the figure, we can see that a value of $k = 6$ results in the highest score on the silhouette and the score is quite reasonable (0.743). Let us visually explore the clusters resulting

**Fig. 5.8** Silhouette score of k-means for different values of k

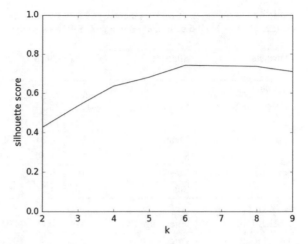

**Fig. 5.9** Visualization of the clusters found with k-means with $k = 6$ (colors) and the labels (markers)

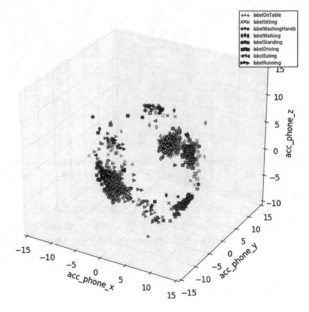

from this setting. We have illustrated them in Fig. 5.9. We also included the labels in the figure. Upon inspection we see quite a nice consistent clustering.

When considering the labels, it does seem that labels are not just randomly spread across the different clusters, for example the yellow cluster seems to contain a lot of walking label points. Table 5.1 shows more statistics about the clusters. We can see that each cluster has its own "niche" in terms of the different accelerometer measurements and we also observe the spread of labels across the clusters. For example, the instances with the phone lying on the table are nearly entirely covered by the first cluster, while the sitting behavior is caught in the third cluster. Walking,

**Table 5.1** Distribution of measurements and labels over for k-mean clustering. Note that the percentage for the label indicates the percentage of total rows among which the label has been assigned

| Attribute | Statistic | Cluster 1 | Cluster 2 | Cluster 3 | Cluster 4 | Cluster 5 | Cluster 6 |
|---|---|---|---|---|---|---|---|
| *Accelerometer data* | | | | | | | |
| acc_phone_x | Mean | −0.36 | 8.24 | 8.00 | −0.75 | −8.21 | −0.56 |
| | Std | 1.15 | 0.96 | 1.02 | 1.71 | 0.94 | 1.21 |
| acc_phone_y | Mean | 0.98 | 0.61 | −2.35 | −9.61 | 2.25 | 9.55 |
| | Std | 1.92 | 1.36 | 2.06 | 1.22 | 1.94 | 1.12 |
| acc_phone_z | Mean | 9.19 | 4.54 | −4.80 | 0.23 | −4.67 | −0.56 |
| | Std | 1.06 | 1.36 | 1.08 | 1.49 | 1.44 | 1.62 |
| *Labels* | | | | | | | |
| labelOnTable | Percentage (%) | 99.56 | 0.44 | 0.00 | 0.00 | 0.00 | 0.00 |
| labelSitting | Percentage (%) | 2.40 | 0.40 | 97.20 | 0.00 | 0.00 | 0.00 |
| labelWashingHands | Percentage (%) | 7.02 | 1.75 | 1.75 | 56.14 | 0.00 | 33.33 |
| labelWalking | Percentage (%) | 1.87 | 0.94 | 0.47 | 46.14 | 0.47 | 50.12 |
| labelStanding | Percentage (%) | 4.74 | 1.42 | 0.00 | 48.34 | 0.47 | 45.02 |
| labelDriving | Percentage (%) | 1.67 | 55.56 | 20.00 | 0.00 | 22.22 | 0.56 |
| labelEating | Percentage (%) | 2.54 | 39.59 | 0.00 | 0.00 | 57.36 | 0.51 |
| labelRunning | Percentage (%) | 17.43 | 0.92 | 17.43 | 64.22 | 0.00 | 0.00 |

standing, and washing hands seem to occur equally frequent among clusters four and six. Having said that, we can expect that this clustering should be able to help us with our tasks.

As a final analysis, we plot the silhouette for the clusters that result per data point, this gives a nice indication on the quality of the various clusters, see Fig. 5.10. We see quite a consistent picture across the clusters.

### 5.8.1.2  K-Medoids

We have applied the k-medoids algorithm to the same problem, and study the same characteristics. In terms of the silhouette scores over different values for $k$ we see a similar result as we have found for k-means: $k = 6$ is best again (Fig. 5.11). The best silhouette score is 0.742, similar to the one we previously obtained for k-means. The clusters are also pretty similar and so are the silhouette scores of the individual points. We therefore only show the table with the statistics, see Table 5.2. The table also does not indicate any significant differences.

**Fig. 5.10**  Silhouette score
of the data points in the
different clusters with $k = 6$

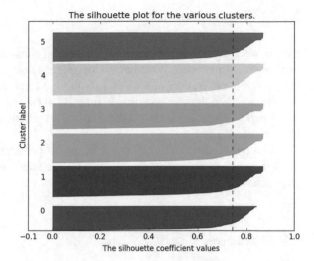

**Table 5.2**  Distribution of measurements and labels over clusters for k-medoids clustering. Note
that the percentage for the label indicates the percentage of total rows among which the labels has
been assigned

| Attribute | Statistic | Cluster 1 | Cluster 2 | Cluster 3 | Cluster 4 | Cluster 5 | Cluster 6 |
|---|---|---|---|---|---|---|---|
| *Accelerometer data* | | | | | | | |
| acc_phone_x | Mean | −0.31 | 8.24 | 7.97 | −0.86 | −8.21 | −0.56 |
| | Std | 1.06 | 0.96 | 1.01 | 1.62 | 0.94 | 1.21 |
| acc_phone_y | Mean | 1.03 | 0.59 | −2.46 | −9.52 | 2.25 | 9.55 |
| | Std | 1.89 | 1.40 | 2.25 | 1.46 | 1.94 | 1.12 |
| acc_phone_z | Mean | 9.26 | 4.53 | −4.76 | 0.30 | −4.67 | −0.56 |
| | Std | 0.87 | 1.38 | 1.08 | 1.54 | 1.44 | 1.62 |
| *Labels* | | | | | | | |
| labelOnTable | Percentage (%) | 99.56 | 0.44 | 0.00 | 0.00 | 0.00 | 0.00 |
| labelSitting | Percentage (%) | 2.40 | 0.40 | 97.20 | 0.00 | 0.00 | 0.00 |
| labelWashingHands | Percentage (%) | 7.02 | 1.75 | 1.75 | 56.14 | 0.00 | 33.33 |
| labelWalking | Percentage (%) | 1.87 | 0.94 | 0.47 | 46.14 | 0.47 | 50.12 |
| labelStanding | Percentage (%) | 4.74 | 1.42 | 0.00 | 48.34 | 0.47 | 45.02 |
| labelDriving | Percentage (%) | 1.67 | 55.56 | 20.00 | 0.00 | 22.22 | 0.56 |
| labelEating | Percentage (%) | 2.54 | 39.59 | 0.00 | 0.00 | 57.36 | 0.51 |
| labelRunning | Percentage (%) | 8.26 | 0.92 | 21.10 | 69.72 | 0.00 | 0.00 |

## 5.8.2   Hierarchical Clustering

Finally, we are going to try a form of hierarchical clustering, namely the agglomerative clustering approach. While we no longer need a pre-defined number of clusters, we can select the number of clusters by choosing a certain point in the dendrogram. We use the Ward linkage function. Figure 5.12 shows the dendrogram for our problem at hand. When we create clusters, the highest silhouette score we obtain is 0.730 for $k = 6$.

Based on the silhouette (which hardly differs per approach) we arbitrarily select the k-means clustering approach to add a feature which represents the attribution of an instance in the dataset to a cluster.

## 5.9   Exercises

### 5.9.1   Pen and Paper

1. We have presented a number of person level distance metrics. While we did not discuss it explicitly, each one comes with their own pros and cons. Present an advantage and a disadvantage for each of the metric (hint: think of computational complexity, the amount of information taken into account, etc.).
2. Let us consider one of the person level distance metrics, namely the dynamic time warping. As we explained the approach, we tried to find the shortest path to match up all of our data points. This would be the distance between the attribute value of two persons. From the literature, it is known that the shortest path is not always the best option to use as a distance metric. Give an example that supports

**Fig. 5.11**  Silhouette score of k-medoids for different values of k

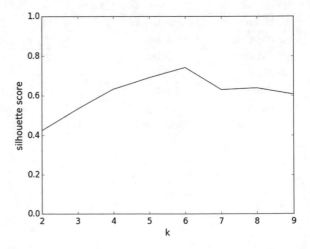

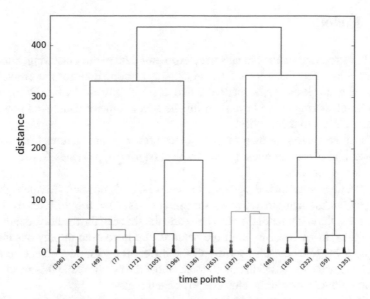

**Fig. 5.12** Dendrogram for the crowdsignals dataset. Note that the numbers between brackets at the *bottom* of the dendrogram represent the number of instances

this statement. Furthermore, give an alternative dynamic time warping distance metric that avoids this disadvantage.

3. We have explained the k-mean and k-medoids algorithms. We did not talk about the guarantees that are provided on the quality of the solution though. Do you think we are guaranteed to find the optimal clustering in either k-means and k-medoids clustering? Explain why (not).

4. The literature suggests that it is very hard to use k-means clustering in combination with some of the person level distance metrics. Which one of the distance metrics would be hardest to use in combination with k-means? And why?

5. What is the computational complexity of both the k-means and the k-medoids algorithm?

6. In agglomerative clustering, Ward's criterion can result in a very different dendrograms compared to the other criteria such as the single and complete linkage criteria. What would you expect to be different if you compare the outcomes of the two approaches? And what could be the reason for this?

7. We have not used a subspace clustering approach in our case study but consider it to be a very relevant approach for the quantified self domain. Provide a concrete example (in terms of a dataset with certain features and certain distributions of the values of those features) where subspace clustering would be a far better choice compared to the more common clustering approach.

8. We have seen one metric to evaluate the quality of clustering namely the silhouette. Provide at least one additional metric for clustering quality and explain how it is computed.

## 5.9.2   *Coding*

1. We have focused on the phone's accelerometer data in our clustering, but did not touch upon the other sensors. Cluster the gyroscope data for the crowdsignals dataset using k-means, k-medoids, and hierarchical clustering. Do you see a similar clustering as we have seen for the accelerometer data? And how do the clusters relate to the activity?
2. Let us move on to your own dataset. Select a few relevant features from your own dataset and cluster them using one of the clustering approaches. Write down and illustrate your results.
3. Select either your own dataset, or the crowdsignals dataset. Compare different criteria for the agglomerative clustering and visualize the differences. Explain how the criteria influence the clustering and the shape of the resulting dendrogram.
4. Take the dataset covering multiple persons you have used in previous chapters. Apply k-medoids clustering with all the different person level distance metrics that have been discussed in this chapter. Show the results using these metrics and compare the results of the clustering for each metric.
5. Apply hierarchical clustering to the dataset covering multiple persons. Use only one person level metric (you can select which one). Compare the outcome to the k-medoids clustering result with the same person level metric.

# Chapter 6
# Mathematical Foundations for Supervised Learning

In this chapter we provide a conceptual and mathematical basis for supervised learning. The reasons for focusing on supervised learning for this more theoretical chapter are twofold: first, we will mainly use supervised learning methods for analyzing quantified self data, and, second, understanding the theoretical underpinnings helps to make the right choices in practical applications and to evaluate results. If you want to dive deeper into theory after reading this chapter, we recommend to read the excellent text books by Abu-Mostafa et al. [2], Shalev-Shwartz et al. [105], and Mohri et al. [87].

We discuss the process and elements of learning and important aspects such as model evaluation and selection. Finally, we will summarize the main theoretical ideas for supervised learning tasks.


## 6.1  Learning Process and Elements

What do we mean when we say "machines can learn"? Take one of the examples from Chap. 1: "Predicting the next blood glucose level based on past measurements and activity levels". This example immediately shows two main ingredients of learning: First, a task is described—predicting the blood glucose level. Throughout the book it is often a predictive model we derive from data. These models can then be used to inform or automate decisions such as taking a certain dose of insulin or recommending additional activities. Second, historical data, here past measurements on the temporal evolution of the glucose level and its link to activity, is needed. In our daily life we often refer to this "data" as experience or examples which we can learn from.

A third aspect is not immediately obvious from the blood glucose example and relates to the goal of learning. When we learn something it is not only about memorizing things but, instead, accomplishing tasks better than we would without learning.

© Springer International Publishing AG 2018

M. Hoogendoorn and B. Funk, *Machine Learning for the Quantified Self*,
Cognitive Systems Monographs 35, https://doi.org/10.1007/978-3-319-66308-1_6

To judge this, it needs a performance measure that tells us how good we are doing on a task, e.g. the mean absolute error when predicting the blood glucose level based on recommended activities. Taking these three constituents (task, experience, and performance measure) we follow Tom Mitchell (1998) in his seminal book [85] and define machine learning as:

**Definition 6.1** A computer program is said to learn from experience E with respect to some class of tasks T and performance P, if its performance at tasks in T improves with E.

This definition extends Definition 1.3 from the first chapter in two ways. First, identifying patterns is not the ultimate objective but an intermediate step towards achieving a task. Second, we need a way to evaluate the performance of learning. This might be as simple as indicated in Defintion 1.6 where we try to estimate a function and thus minimize an error, or more complicated as in the case where for evaluating an action that tries to accomplish task $T$. Choosing an appropriate performance measure is a very important step in machine learning, because it heavily impacts the learning behavior of the learning algorithm. At the same time, as you will see, the choice is not easy and heavily depends on the task itself.

## 6.1.1   Unknown Target Function

We will now take a closer look on what computers need in order to learn and how learning is achieved. Let us focus on supervised learning (Definition 1.6). The assumption behind supervised learning is that many real-world phenomena can be described in terms of a mathematical model. A model simplifies the perspective on a real-world phenomenon by the abstraction of, and focusing on, its important aspects. It refers to a functional input/output relationship of observable concepts. Think a little bit about the model oriented perspective—it is very powerful not only for understanding real-world phenomena but also for predicting them. Let us see what this means by reconsidering the blood glucose example. Many factors influencing the blood glucose level have been studied in detail: So, for example, it is known, that stress, fat, coffee, and medication cause the glucose level to increase, while sports and healthy food can decrease it. In addition, the functional relationship will be specific to individuals as you might expect from our two characters Arnold and Bruce. Knowing the functional relationship for our two guys enables us to understand how their individual glucose levels evolve and also to come up with personalized recommendations for training and food.

The main ingredient here is the *functional relationship* that we would like to learn. As such, machine learning has many similarities with inferential statistics but extends its capabilities. We call the functional relationship that we want to learn, the *unknown target function*[1] $f$ (see Fig. 6.1). It quantitatively represents the model and can be

---

[1] In machine learning the terms function and model are often used interchangeably.

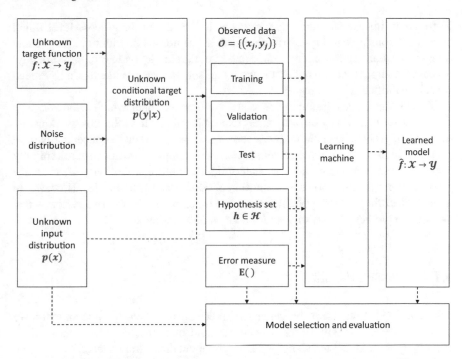

**Fig. 6.1** Process and elements of supervised machine learning (based on the learning diagram from Abu Mostafa et al. 2012 [2])

any type of function $f : \mathcal{X} \to \mathcal{Y}$ that maps the observation $x \in \mathcal{X}$ to the target $y \in \mathcal{Y}$ (note that the argumentation also holds for a categorical target $g$). Here, $\mathcal{X}$ (and $\mathcal{Y}$) represent the input (and target) space which is the set of all potential observations.[2] You can think of the unknown target function as the underlying generator of the data we observe.

In almost all realistic cases, the data will not be generated by a deterministic target function. That is because we often have measurement errors on the input side (e.g. when recording his food intake Bruce unintentionally forgets about the last piece of cream cake he had for breakfast) but also some noise on the output (e.g. we have not accounted for all factors that influence the glucose level). These two sources of noise would need to be dealt with independently, especially in cases of non-linear relations. However, as usual in machine learning, we will restrict ourself and account only for noisy targets by introducing what we call the *unknown conditional target distribution* $p(y|x)$. If you have not seen the notation $p(y|x)$ (read as the probability of $y$ given $x$), do not be scared. It gives the probability distribution of $y$ assuming that $x$ is fixed. So, instead of having a deterministic function, we take an input value $x$, calculate $f(x)$, and then add some noise that is generated from a *noise distribution*. When the target is discrete the noise distribution can be the Bernoulli or categorical

---

[2] For reasons of simplicity, we drop the index $j$ of observations $(x_j, y_j)$ in the following paragraphs.

distribution. In case of a numerical target as in the case of the glucose level it will be a continuous probability distribution such as the normal distribution. As you can imagine, accounting for random noise makes it more tricky (at least approximately) to learn the target function but that is what realistic datasets are like as we have seen for the quantified self in Chap. 3.

There is one piece missing to "generate" the observed data, which is the *unknown input distribution* $p(x)$—as you might expect it tells us how the observations are distributed. We will see later in this section why it is important to conceptually have $p(x)$ as a building block of our learning setup even though we do not need to explicitly know it. Note, that we could now determine the joint distribution $p(x, y)$, since $p(x, y) = p(y|x) \, p(x)$ holds by definition of conditional probabilities. However, in machine learning we are typically interested in learning $p(y|x)$. The reason is that we want to predict a target value $y$ for a given observation $x$.

### 6.1.2   Observed Data

$p(x)$ and $p(y|x)$ represent the generating mechanism behind the phenomenon we study. However, the only thing we can get hold of is the observed dataset $O = \{(x_j, y_j) | j = 1, ..., N\}$ which helps us to learn about the unknown target distribution $p(y|x)$. Thus, the observed data can be interpreted as a blurry filter through which we see the theoretical model.

Depending on the learning algorithm we split the observed data in two or three subsets. The first subset is called the *training dataset*. It is the main piece of data used to learn which function $\hat{f}(x)$ fits the observed data best. Depending on the amount of data we are given the fraction of the overall dataset that constitutes the training dataset varies. When data is limited, as it is often the case in psychological or medical research, around two third of the overall dataset is used for training purposes. When there is plenty of observed data available another aspect becomes important. Assume you have a binary classification problem at hand (e.g. want to predict a critical state of the blood glucose level): Very often, the number of observations belonging to one of the two classes is heavily skewed. We say the dataset is *unbalanced*. If we would present this to our learning algorithms the learned target distribution would put a strong weight on the dominant class in the data. In fact, it might achieve a very good performance by doing so. However, we would be interested in predicting the underrepresented class in a good way as well. For instance, if we want to predict a rare disease we are most interested in a model that is able to identify the people that might have or develop the disease. In this case, to facilitate learning, the training data is designed to be a *stratified* sample of the overall observed data. That means, training examples are chosen from the observed data in a way that both class labels are equally likely, which then directs the optimization algorithm to pay more attention to the underrepresented class.

Once the best fitting function is found, the *test dataset* (our second subset) helps to evaluate whether $\hat{f}(x)$ is also a good fit in a sense that it generalizes well—that is, calculating the target value for unseen data based on $\hat{f}(x)$ will produce good predictions (what "good" means depends on how and what for $\hat{f}(x)$ is used in practice; this issue will be discussed in more detail in Sect. 6.1.5). It is important that the test dataset is kept separate from the learning process and is only used to assess its generalizability.

In supervised learning we always have a training and a test dataset. In addition to that some model types (e.g. decision trees) use a *validation dataset* that advises the learning process when to stop. This can also help to tune the parameter values of the learning algorithms. We will explore this in more detail when discussing decision trees (Sect. 7.5) but you can think of the validation data as being used to set a stopping criteria for learning and thus to avoid *overfitting* (see Sect. 6.2).

For a small dataset we have to use the limited data in the best possible way. One approach is to use cross validation: we split our data into $k$ chunks (without any overlap in the folds), use $k - 1$ chunks to train on, and one chunk to test upon. We have $k$ different configurations for the training and the test set then. This is called $k$-fold cross validation. We average the performance over these $k$ configurations. We fully use the data in such a case, but never train on data we test upon.

### 6.1.3 Error Measure

We know that in general we are not able to infer the exact target function $f$. This is due to the noisy target distribution and the finite amount of observed data available for learning. To understand how far apart a candidate function $h$ is from the unknown target function $f$ we need an error measure $E(f, h)$ that takes the two function as arguments and returns a real value. To make this operational let us use a point-wise error $e(f(x), h(x))$, that is a distance measure between the two values $f(x)$ and $h(x)$. Based on that we define $E(f, h)$ to be the expected value of $e(f(x), h(x))$ given $p(x)$.

$$E(f, h) = \int_{x} e(f(x), h(x))p(x)dx \tag{6.1}$$

In the more theoretically oriented literature (e.g. [121]) $e(f(x), h(x))$ is called *loss* and $E(f, h)$ is called *risk* or *risk functional*.

Now that we have defined a general error measure, can we calculate it easily? As you probably have already realized, the answer is "no" for any realistic learning problem. Why? In order to calculate $E(f, h)$ we would not only need to know $h(x)$ at every point in the input space but also $f(x)$. However, if we could determine $f(x)$ at any point, it would not be *unknown* anymore and there would not be anything to learn. So what can we do instead? We approximate $E(f, h)$ as follows:

$$E(f, h) \approx \frac{1}{N} \sum_{j=1}^{N} e(y_j, h(x_j)) \tag{6.2}$$

The term on the right hand side is called empirical risk and plays an important role, when it comes to learning $h(x)$. We will see later in this chapter that this approximation can be reasonably good but also go utterly wrong if we are not careful when choosing the hypothesis set. For now, let us consider a classification problem and see what Eq. 6.2 looks like. One of the obvious error measures is the rate of correct classification, that is the fraction of cases we correctly classify using our learned model:

$$\text{classification rate} = \frac{|\{g_j \in \mathbf{G} | \hat{g}_j = g_j\}|}{N} \tag{6.3}$$

We can also be a bit more precise to see where we make mistakes. Consider Table 6.1, also known as a confusion matrix, and let us assume we have a binary classification problem (i.e. we need to predict 0 or 1). In the table, the columns represent the predictions of the model $(\hat{g}_j)$, and the rows are the actual values $(g_j)$. We see that we can be correct in two ways: we predict a positive case (i.e. a 1) correctly (a true positive) or we predict a negative case right (a true negative). Furthermore, if we make a mistake where we predict a positive but it is actually a negative we refer to it as a false positive. Similarly we have the concept of a false negative.

Based on these concepts, we can define additional measurements, such as the *precision* and *recall*. The precision represents how correct we are when we predict a 1 with our model ($precision = \frac{TP}{TP+FP}$). The recall measures how many of the positive cases we identify based on our model ($recall = \frac{TP}{TP+FN}$).

Ideally, we would like to perform well on both precision and recall. Therefore, the so-called F-measure that combines the two has been developed:

$$F_\beta = \frac{(1 + \beta^2) \cdot \text{precision} \cdot \text{recall}}{\beta^2 \cdot \text{precision} + \text{recall}} \tag{6.4}$$

The parameter $\beta$ expresses how the precision and recall are weighted. It attaches $\beta$ times as much importance on recall compared to precision.

We will see in the next section how the regression case can be handled in terms of error measures. But before doing so, we want to combine what we have learned

**Table 6.1** Confusion matrix

| Actual | Predicted | |
|---|---|---|
| | 1 | 0 |
| 1 | True positive (TP) = $\frac{\{g_j \in \mathbf{G} | \hat{g}_j = g_j \wedge \hat{g}_j = 1\}}{N}$ | False negative (FN) $\frac{\{g_j \in \mathbf{G} | \hat{g}_j \neq g_j \wedge g_j = 1\}}{N}$ |
| 0 | False (FP) $\frac{\{g_j \in \mathbf{G} | \hat{g}_j \neq g_j \wedge g_j = 0\}}{N}$ | True negative (TN) = $\frac{\{g_j \in \mathbf{G} | \hat{g}_j = g_j \wedge \hat{g}_j = 0\}}{N}$ |

about the various subsets of observed data and the error measures. In practice you will find two terms being used often: in-sample error and out-of-sample error. The concept behind is as simple as it is important. The term "sample" refers to the chunk of observed data we called training data. Based on Eq. 6.2 the in-sample error is defined as:

$$E_{in}(h) = \frac{1}{N} \sum_{(x,y) \in \mathcal{O}_{Train}} e(y, h(x)) \tag{6.5}$$

Compared to Eq. 6.2 we dropped the reference to the unknown target function $f$ since we use the observed data instead. Thus, the in-sample error $E_{in}(h)$ is the empirical risk calculated for the training data. It is an import measure that will steer the learning process in Sect. 6.1.4. You already guess what the out-of-sample error refers to: Strictly speaking we define it as

$$E_{out}(h) = \int_{\mathcal{X} \setminus \mathcal{O}_{Train}} e(f(x), h(x)) p(x) dx \tag{6.6}$$

$E_{out}(h)$ tells us how well the hypothesis $h(x)$ is doing on the full input space (accounting for $p(x)$) except for the training data. Note that for all practical cases, due to the finiteness of $\mathcal{O}_{Train}$ we have $E_{out}(h) = E(f, h)$. As we will discuss in more detail, learning is about minimizing $E_{out}(h)$ without knowing it exactly. Again, if we could exactly determine $E_{out}(h)$ we would need to know $f(x)$ which would mean that we do not have anything to learn. Therefore we will take $E_{in}(h)$ as a proxy for $E_{out}(h)$ in the next section.

We would like to conclude this section by emphasizing how important it is to choose an appropriate error measure. That is because the error measure strongly impacts the outcome and feasibility of learning. In practice the choice of the error measure depends on the intended use of the model that shall be learned. For example, thinking of our friend Bruce, we might especially be interested to predict extreme low or high values of the blood glucose level with the highest possible precision, while we do not care to be very exact in the mid range. Often, in a first step, a computationally convenient error measure is chosen (e.g. mean squared error, classification rate) and then in a second step adjusted to the needs of a practical application (see Sect. 6.1.5).

### 6.1.4 Hypothesis Set and the Learning Machine

We now focus on another element in Fig. 6.1: the hypothesis set $\mathcal{H}$ which is a set of functions that contains all potential hypothesis $h$ we consider to be candidates to fit the unknown target function well. Very often $\mathcal{H}$ is an infinite set of functions. For example, in regression this could be the (simple yet infinite) set of all linear functions on the input space $\mathcal{X}$.

Thus, learning comes down to the task of selecting the hypothesis $h \in \mathcal{H}$ that *best* matches the observed data—we call this hypothesis $\hat{f}$. You might think that this is a very restrictive approach to learning because we need to have a "good" candidate in our hypothesis set, and in a sense you are right. However, keep in mind, that we could choose $\mathcal{H}$ to be as flexible as we wish to, that is we can add whatever class of function we believe might be appropriate. But adding flexibility comes at a price, it makes the selection of $\hat{f}$ harder, since you have more to choose from with the same amount of observed data. In fact, as discussed for example by Mitchell [85], without restricting the hypothesis space, learning as a selection of a $\hat{f}$ from $\mathcal{H}$ is not possible.

We have everything in place (in particular concepts such as observed data, error measure, and hypothesis set) to tackle the actual learning as represented by the *learning machine* in Fig. 6.1. We said that learning is the task of selecting the hypothesis $h \in \mathcal{H}$ that *best* matches the observed data. What we mean by *best* is, that we choose $h$ so that it minimizes the in-sample error

$$\hat{f} = \operatorname*{argmin}_{h \in \mathcal{H}} E_{in}(h) \tag{6.7}$$

So, behind the *learning machine* is an optimizer that tries to solve the above problem. Since we assume you are familiar with linear regression from an introductory class on statistics we take it as an example here: What is the hypothesis set in the linear regression case? We define an individual hypothesis as follows:

$$h_\theta(x_j) = \theta_0 + \sum_{k=1}^{p} \theta_k x_j^k = \theta^T x_j \tag{6.8}$$

where $\theta$ is the unknown vector of parameters and $\theta_k$ are its components (the subscript of $h_\theta(x_j)$ denotes the dependency on $\theta$). Note that the vector $\theta$ has a length of $p+1$ (the subscript runs from 0 to $p$), while an observation $x_j$ as defined in Eq. 3.1 only has length $p$. A common trick is to extend the $x_j$ by adding a constant 0-th element $x_j^0 = 1$ that links to $\theta_0$ which in linear regression is called the intercept.

What are we missing? In order to minimize the in-sample error, we need to fix the point-wise error. A convenient choice is $e(y, h(x)) = (y - h(x))^2$ which leads to what you probably already know as the least-mean-square solution in linear regression. Applying Eq. 6.5 to our specific settings and the full set of observed data leads to

$$E_{in}(h) = \frac{1}{N} \sum_{j=1}^{N} (y_j - \theta^T x_j)^2 = \frac{1}{N} (\mathbf{Y} - \theta^T \mathbf{X})^T (\mathbf{Y} - \theta^T \mathbf{X}) \tag{6.9}$$

What does the in-sample error look like? To get a graphical intuition let us assume that the input space is 1-dimensional and the true (and unknown) parameters that generated the data are $\theta_{True} = (3, 2.5)$ and $N = 200$. As you can see in Fig. 6.2 the in-sample error is a quadratic function that only has one minimum. It is this minimum, referred to as $\hat{\theta}$ corresponding to $\hat{f}$, we are looking for. If you are familiar with linear

**Fig. 6.2** $E_{in}(h)$ is the in-sample error calculated at $\theta$ given the observed data

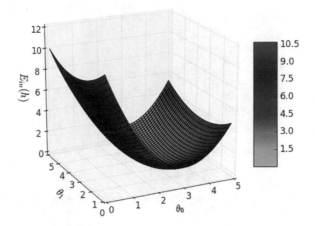

regression you might remember that, under the condition $\mathbf{X}^T\mathbf{X}$ is invertible, there is a closed form solution

$$\hat{\theta} = (\mathbf{X}^T\mathbf{X})^{-1}\mathbf{X}^T\mathbf{Y} \tag{6.10}$$

To minimize the in-sample error we can also use an algorithm that is often used in machine learning: *gradient descent*. The story behind it is simple: Assume you virtually jump into Fig. 6.2 at an arbitrary position, say $\theta_0 = \theta_1 = 0$, and then go down-hill step by step. Eventually, you will arrive at $\hat{\theta}$ which represents the minimum of the in-sample error. So, (machine) learning is like white water rafting and not as is often said "swimming upstream".

How can we technically implement gradient descent? The gradient of a multi-parameter real-valued function is the vector of the partial derivatives of that function. It points into the direction of the steepest ascent (see left panel in Fig. 6.3). If we want to determine the next step, we can apply the following rule:

$$\theta_k = \theta_k - \eta\frac{\partial}{\partial\theta_k}E_{in}(h) \tag{6.11}$$

where $\eta$ is the so-called learning rate that determines whether and how fast the gradient descent algorithm converges. It is not too hard to show (see exercise) that the update rule can be formulated in a vectorized notation:

$$\theta = \theta - \eta\nabla_\theta E_{in}(h) = \theta - \frac{\eta}{N}\mathbf{X}^T(\mathbf{X}\theta - \mathbf{Y}) \tag{6.12}$$

where $\nabla_\theta$ is the so-called nabla operator that calculates the gradient of the function it is applied to.

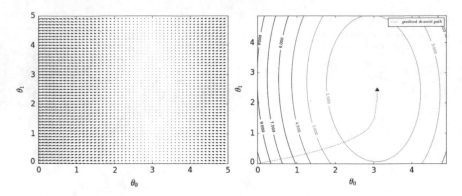

**Fig. 6.3** *Left panel* shows the gradient field, *right panel* shows gradient descent for the same data

Equation 6.12 is often referred to as the Widrow-Hoff algorithm. It makes intuitively sense that it contains a term $(\mathbf{X}\theta - \mathbf{Y})$ representing the deviation of the hypothesis $h_\theta(x)$ from the observed $\mathbf{Y}$. The larger this difference is, the faster we go down-hill. We implement the Widrow-Hoff algorithm or, as it is also called, the "least-mean-square (LMS)" training rule. The step-wise gradient descent is shown in the right panel of Fig. 6.3. The implemented version of the algorithm is called *batch gradient descent* because all training examples are used at the same time. There are other variants of the gradient descent algorithm that update $\theta$ by using individual training examples separately (*stochastic gradient descent*) that are useful in situations where new data comes in.

We started with the closed form solution of the linear regression case. Compared to the Widrow-Hoff algorithm it looks much simpler and more elegant. However, there are two main reasons why the iterative approach is very powerful. First, Eq. 6.10 involves calculating the inverse of $\mathbf{X}^T\mathbf{X}$ which can be computationally expensive given its size $(p+1)^2$ for complex problems. In contrast, the Widrow-Hoff algorithm does not need to calculate the inverse but only needs matrix multiplications. Second, the closed form solution is closely coupled to the mean squared error we used for the point-wise error $e(y, h(x))$. When we need to change the point-wise error, in most cases there will not be a closed form solution anymore. On the other side, the pre-condition for applying gradient descent is only that the point-wise error is differentiable. When discussing neural networks we will see that this will help to find solutions.

To conclude this section, we want to emphasize two important aspects: First, this section touches the core of machine learning since it demonstrates how machines can learn in the sense of picking the best hypothesis from an infinite set of hypotheses. In Chap. 7 we will see other learning algorithms such as the perceptron learning rule. However, the basic idea of minimizing $E_{in}(h)$ will stay.

Second, learning requires to choose a specific (and limited) hypothesis set before running the learning algorithm. If, instead, we would define the hypothesis set to be the set of all functions that maps $\mathcal{X}$ to $\mathcal{Y}$ the learning process described above would not work. For sure, it would be easy to find many functions that minimize $E_{in}(h)$. In fact you can construct an infinite number of functions, that yield $E_{in}(h) = 0$. But, in general, they would overfit the unknown target function. We will discuss this in more detail in Sect. 6.2.

### 6.1.5 Model Selection and Evaluation

This section is concerned with two questions: First, how can we select the best model, that is, finding the model which will perform best in a given task—e.g. predicting Bruce's mood this evening? Second, how can we assess the quality of a model we have learned—e.g. how often will we be right when predicting Bruce's mood in the future?

In Sect. 6.1.4 we have shown that minimizing $E_{in}(h)$ is the way how machines learn. As a result of this optimization task we are left with *one* function $\hat{f}$ that is supposed to approximate the unknown target function. Having said that, the first question we posed at the beginning of this section seems a bit odd—we have one $\hat{f}$, so there is not too much of a choice, right? In the following we will discuss three cases where model selection is yet an issue.

Let us have a closer look at binary classification problems. Instead of taking linear combinations of all input variables as our hypothesis set as in the regression case (Eq. 6.8) we define our hypothesis set as:

$$h_\theta(x_j) = \begin{cases} 1, & \text{if } \Theta_{cut} < (1 + e^{-\theta^T x_j})^{-1} \\ 0, & \text{otherwise.} \end{cases} \tag{6.13}$$

where $\Theta_{cut}$ is a threshold parameter that determines how restrictive the hypothesis $h_\theta(x_j)$ acts when allocating a sample to class 1. Depending on the error measure, that we have not defined so far, we could apply the methodology of the previous chapter to estimate the parameters $\theta$ and $\Theta_{cut}$ simultaneously. On the other hand, as you might have realized, Eq. 6.13 represents the well-known logistic regression and parameters $\theta$ can be learned using Maximum Likelihood. This is done by interpreting $(1 + e^{-\theta^T x_j})^{-1}$ as the probability that $x_j$ belongs to class 1 (see also the exercise about this). Hence, we might want to select $\Theta_{cut}$ after this learning process.

Now, assuming that we want to classify an unseen $x$, we would get a scalar value $0 \le \theta^T x \le 1$ that we could interpret as a probability. Depending on $\Theta_{cut}$ we can predict to which class $x$ belongs most likely. If we vary $\Theta_{cut}$ we obtain the so-called Receiver Operating Characteristic (ROC for short) curve (Fig. 6.4).

The ROC curve shows the trade-off between the fraction of true positives that are found out of the total number of positives versus the fraction of false positives that were found among the total number of negatives. In Fig. 6.4, the dashed line represents a random classifier while the solid line is an example performance of a better classifier. Ultimately we would like the curve to go straight to the upper

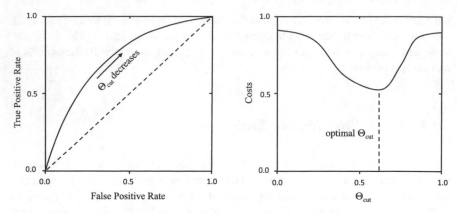

**Fig. 6.4** *Left panel* shows the ROC curve, *right panel* displays the dependency of costs associated with using the classification model

left corner: this would mean that we are able to identify all positive cases without misclassifying a single actual negative case as positive. The measure of performance used with respect to the ROC curve is often the Area Under the ROC curve. A value of 0.5 equals random while a value of 1 (i.e. the line all the way to the upper left corner that was discussed) is optimal.

Model selection for the above situation means choosing an appropriate $\Theta_{cut}$. How do we do that? Let us turn to Bruce's blood glucose level again and assume that we are interested in predicting whether it is within a healthy range or not. We want to use these predictions to come up with recommendations for Bruce: if the level is out of range, Bruce is asked to "check his blood sugar", otherwise "no action is required". The prediction and derived recommendation can be wrong in two ways: First, the glucose level might be ok, but our model suggests that is not. In this case, Bruce would be asked to check his blood sugar and could relax after the false alarm. The effort and costs for the (un-needed) test are reasonably low. Second, the glucose level is out of range, but our model does not recognize it. Bruce might run into a life threatening situation which we could have prevented, if our model would have been right. It is obvious that mistakes of the second type are much more harmful, so we would like to avoid them even though we might make more mistakes of the first type.

The ratio of mistakes is influenced by $\Theta_{cut}$. As we vary $\Theta_{cut}$ the numbers in Table 6.1 and thus the position on the ROC curve change (left panel Fig. 6.4). In order to determine the optimal $\Theta_{cut}$ we can use this dependency. To do so, we have to quantify the cost or effort associated with the four fields in Table 6.1, which in practice can be very tricky (e.g. what is cost of "life threatening"?). Once we are given the "cost matrix" corresponding to Table 6.1 we can calculate the expected cost for applying the classification model. The right panel of Fig. 6.4 shows that an optimal $\Theta_{cut}$ can be picked. Having now both, an optimal $\theta$ and $\Theta_{cut}$, we have selected the best performing model.

We said at the beginning of this section that we want to explore three different cases in which model selection plays a role, the first was covered in the previous paragraphs, let us look at another common situation which requires model selection: Today, we are often faced with large amounts of data where large refers not only to the number of samples but also the available attributes we can use for learning. As we have seen in Sect. 4.3 when analyzing texts we quickly have to deal with high-dimensional input spaces—think of a word as defining a dimension. However, high dimensionality can also occur, when dealing with seemingly low dimensional data as for example a sequence from an accelerometer sensor. How so? Instead of taking the raw data from the sensor as input for machine learning, we have seen that the raw data is aggregated on different time scales and other measures such as periodicity are used as input. This can be done in an infinite number of ways.

So, we might end up with having millions of features we could use for learning. What happens if we apply the empirical risk minimization approach from Sect. 6.1.4 to a situation where $p > N$ (see exercise to explore this in more detail)? As you might remember from linear regression there is no single best solution (since $X^T X$ is not invertible). It was the Russian mathematician Andrey Tikhonov who recognized that a unique solution exists when instead of taking the empirical risk only (remember that is what we also called the in-sample error $E_{in}(h)$) we add a penalization term to the empirical risk and minimize the sum of both. This approach is called *regularization* and we will learn more about it in Sect. 7.9.2. For now, we only need to know that there is a positive regularization parameter $\lambda$ that governs how many features will be included in the regression. To determine the optimal $\lambda$ that (presumably) minimizes $E_{out}(h)$ we learn a set of models $h$ by varying $\lambda$. Which model do we ultimately select from this set to be our optimal model? While the learning in the previous step is done using only the training data, we now calculate for each model the corresponding $E_{in}(h)$ on the validation data. The model that has the lowest in-sample error on the validation data gets selected as our best model.

Let us consider the third case where model selection is necessary. So far, we have assumed to have a fixed hypothesis set $\mathcal{H}$. In practice that might be true and you might have some prior knowledge that determines $\mathcal{H}$. However, in most cases $\mathcal{H}$ cannot be derived from first principles. For example, we might not only consider linear combinations of all features as $\mathcal{H}$ but also high order polynomials. Let $\mathcal{H}_d$ denote the set of multivariable polynomial of degree $d$, then we know that $\mathcal{H}_1 \subset \mathcal{H}_2 \subset \mathcal{H}_3 \ldots$. Therefore, with increasing degree $d$ the hypothesis sets get more powerful in capturing complex relations of the model which translates into a decreasing in-sample error. As a direct result the following inequalities hold

$$\min_{h \in \mathcal{H}_1} E_{in}(h) \geq \min_{h \in \mathcal{H}_2} E_{in}(h) \geq \min_{h \in \mathcal{H}_3} E_{in}(h) \ldots \tag{6.14}$$

In fact, no later than $d \geq N - 1$ the in-sample error will be 0, we refer to this situation as overfitting. The term overfitting comes from the fact that we perfectly fit our training data, while the out-of-sample error is very likely to even increase.

How can we avoid overfitting? We apply the same procedure as before: For each hypothesis set $\mathcal{H}_d$ we minimize the in-sample error with respect to the training data and obtain the best fitting model $h_d$. For each $h_d$ we then calculate the in-sample error with respect to the validation data. Ultimately, we choose $h_d$ with smallest error in the validation data as our best model, thereby finding the best overall model.

Up to this point we have not touched the test data, neither for the training nor the model selection. Therefore we can now evaluate our model by taking its in-sample error on the test data as a proxy for the out-of-sample error. It cannot be overemphasized how important this strict separation of the test data from training or validation data is, only then will the assessment of the out-of-sample error be reliable.

## 6.2   Learning Theory

In the previous section we studied the learning process and its elements. The question we are concerned with in this section is: is it guaranteed that machines can learn as discussed above? The answer of Radio Yerevan would probably be: "In principle, yes, but ...". To elaborate this answer, we present in this section the main ideas of the underlying theory without going into the mathematical details. Our goal is to provide a conceptual underpinning for choosing the right hypothesis set and evaluating the learning results. You might find it tricky to understand every step in this section, if you do not have some working knowledge in probability theory. In this case, try to follow the basic flow of arguments.

### 6.2.1   PAC Learnability

Let us rephrase and sharpen the initial question of this section: *Can machines always learn the unknown target function?* We have seen in the previous section that we need lots of training examples to be able to approximate the unknown target function $f$ well. The question is: Will we ultimately succeed in approximating the $f$ arbitrarily well, when the number of training samples increase ($N \to \infty$)? Perhaps surprising, but given the setting in the previous section, the answer is in most cases No! This is because an obvious prerequisite, i.e. $f \in \mathcal{H}$, is typically not guaranteed in practical applications. Therefore, no amount of training examples will be sufficient to nail down $f$.

So, let us try to address the initial question of this section differently. We say learning is possible, if the in-sample error is a good estimator for the out-of-sample error: the in-sample error might be and in all relevant cases is not 0. However, with increasing $N$ the difference $|E_{out}(\hat{f}) - E_{in}(\hat{f})|$ will be most likely very small. Why "most likely"? There is always the chance that the random training examples are drawn from $p(x, y)$ in a way that the above difference of errors is large. This is captured in the following definition:

**Definition 6.2**  A hypothesis set is said to be *PAC learnable* (probably approximately correct), if a learning algorithm exists that fulfills the following condition: For every $\epsilon > 0$ and $\delta \in (0, 1)$ there is an $m \in \mathbb{N}$, so that for a random training sample with length larger than $m$, the following inequality holds with probability $1 - \delta$: $|E_{out}(\hat{f}) - E_{in}(\hat{f})| < \epsilon$

Admittedly, this definition looks a bit complicated. Basically we call a hypothesis set PAC learnable, if given enough training examples we can approximate the out-of-sample error arbitrarily well by the in-sample error. The statement comes with a probabilistic caveat (due to the above mentioned random nature of the sampling process), it only holds with a certain probability $1 - \delta > 0$. However, this probability can be chosen to be arbitrarily small.

Let us consider a very simple hypothesis $\mathcal{H}_1$ set that only consists of one function $h$ and we are given $N$ training samples. A learning algorithm would for sure output $h$ (because it is the only element in $\mathcal{H}_1$). Using the famous *Hoeffding's inequality* we have

$$p(|E_{out}(h) - E_{in}(h)| > \epsilon) \le 2e^{-2\epsilon^2 N} \qquad (6.15)$$

That means with increasing $N$ the difference between in- and out-of-sample error will most likely become arbitrarily small (corresponding to whatever we set $\epsilon$ to be). Setting $\delta = 2e^{-2\epsilon^2 N}$, we therefore conclude that this simple hypothesis set is *PAC learnable* as defined in Definition 6.2. Since $\mathcal{H}_1$ is neither a very exciting nor realistic case, we move on to a more complex hypothesis set.

We assume that another hypothesis set $\mathcal{H}_M$ contains a finite number $M$ of hypotheses. Now for every $h \in \mathcal{H}_M$ the Eq. 6.15 holds. However, the learning algorithm will finally pick only one hypothesis $\hat{f}$ that is supposed to minimize the error measure. Therefore if we want to extend Eq. 6.15 to the larger hypothesis set $\mathcal{H}_M$ and bound the error difference for $\hat{f}$, we have to account for the number of choices. Using the *union bound* we get

$$p(|E_{out}(\hat{f}) - E_{in}(\hat{f})| > \epsilon) \le 2Me^{-2\epsilon^2 N} \qquad (6.16)$$

Note that the subtle difference to 6.15 is that the right hand side of the inequality is by a factor of $M$ larger—that is exactly what the union bound says. Even though this bound is less good than Eq. 6.15, we see that the preconditions of Definition 6.2 are still fulfilled and thus, that every finite hypothesis set is PAC learnable. Equation 6.16 can be rewritten

$$E_{out}(\hat{f}) \le E_{in}(\hat{f}) + \sqrt{\frac{1}{2N} \log \frac{2M}{\delta}} \qquad (6.17)$$

which holds with probability $1 - \delta$. The fact that every finite hypothesis set is PAC learnable is already a very powerful result, even though most hypothesis sets used in practice are not finite. Why? Take for example the set of all linear functions for a given number of features. Since we could arbitrarily choose parameters $\theta_k$ in Eq. 6.8 this hypothesis set for sure contains an infinite number of hypotheses. However, we can construct a new hypothesis set for which all $\theta_k$ are bounded and discretized. This yields a hypothesis set that is finite while being similar to the initial set, depending on the granularity of discretization. Thus, using Eq. 6.16 we know that this discretized hypothesis set is PAC learnable.

## 6.2.2   VC-Dimension and VC-Bound

We have seen that finiteness of the hypothesis set is a sufficient condition that it is PAC learnable. In this section we demonstrate that there are also infinite hypothesis sets that are PAC learnable. We start with an example: Assume you have a binary classification problem and only one real-valued input variable $x$. Your infinite hypothesis set $\mathcal{H}_{step}$ shall consist of all step functions

$$h_\theta(x) = \begin{cases} 1 & x \leq \theta \\ 0 & otherwise \end{cases} \tag{6.18}$$

Let the unknown target function be one of the above step functions with an unknown $\theta_{true}$ and no noise. Thus, what we want to learn is the unknown $\theta_{true}$. By intuition you can imagine, that the more random training examples we have, the higher is the chance to get close to $\theta_{true}$. If $\theta_{learn}$ is the output of our learning algorithm, $|\theta_{learn} - \theta_{true}|$ determines the out-of-sample error. Therefore depending on the number of training examples we can get as close as we want to $\theta_{true}$. Thus, given Definition 6.2 the hypothesis set $\mathcal{H}_{step}$ is PAC learnable, even though it is infinite.

To generalize this result to other infinite hypothesis sets we want to better understand the expressive power or complexity of a hypothesis set. By the expressive power of a hypothesis set we mean its ability to correctly represent the training examples. Let us focus on a binary classification problem and define a restriction of a hypothesis set $\mathcal{H}$ to a finite set of input vectors $X$ as follows [105]

**Definition 6.3** Let the hypothesis set $\mathcal{H}$ be a set of functions $h : \mathcal{X} \to \{0, 1\}$. For a set of input vectors $X = \{x_1, x_2, \ldots, x_N\}$, we call $\mathcal{H}_X = \{(h(x_1), \ldots, h(x_N)) : h \in \mathcal{H}\}$ a *restriction of $\mathcal{H}$ on $X$*.

An element of $\mathcal{H}_X$ is build by taking a function $h \in \mathcal{H}$ and applying it to all $N$ vectors in $X$. This yields a vector of length $N$ whose elements are 0's and 1's Thus, $\mathcal{H}_X$ contains a number of such length $N$ vectors. Since there are no more than $2^N$ such vectors, the size of $\mathcal{H}_X$ can maximally be $2^N$. However, the size of $\mathcal{H}_X$ is often smaller than $2^N$, because various $h$ produce the same 0/1-vector. If the size of $\mathcal{H}_X$ is

$2^N$, we say $\mathcal{H}$ shatters $X$, i.e. $\mathcal{H}$ can realize every possible labeling of the $N$ input vectors. Based on that we define:

**Definition 6.4** The Vapnik-Chervonenkis (VC) dimension $d_{VC}$ of a hypothesis set $\mathcal{H}$ is the maximum number of input vectors that can be shattered. The VC dimension is infinite if there are arbitrarily large sets of input vectors that can be shattered.

Note that the VC dimension of a hypothesis set $d_{VC}$ does not impose that any set of input vectors that has size $d_{VC}$ can be shattered. Instead, the definition only requires the existence of at least one set of $d_{VC}$ input vectors that can be shattered.

We admit that this might sound a bit complicated, so let us look at an example: Assume the input space $\mathcal{X}$ is $R^2$ and the hypothesis set $\mathcal{H}$ encompass all affine functions in $R^2$, this is all straight lines. In Fig. 6.5 you see a set of three points in $R^2$. It is possible to find 8 different hypotheses (straight lines) that separate the three points in every of the $2^3$ ways. Can we shatter every set of three points by $\mathcal{H}$? No, think of three points sitting on a straight line. Nevertheless, we have found a set of three points that can be shattered by $\mathcal{H}$, so we know by definition that $d_{VC} \geq 3$.

Can we choose 4 points in a way that they all can be shattered by $\mathcal{H}$? As an exercise convince yourself that this is not possible. Therefore we know that $d_{VC} < 4$ and thus $d_{VC} = 3$.

What can we use the VC dimension for? Being able to shatter a large number of points in $\mathcal{X}$ and thus having a high VC dimension $d_{VC}$ corresponds to a characteristic of a hypothesis set that we call complexity. The higher the complexity of a hypothesis set, the better we are able to model training examples.

Given a hypothesis set $\mathcal{H}$ with an infinite VC dimension and a fixed set of $N$ training examples it is likely that there is a hypothesis $\hat{f} \in \mathcal{H}$ that perfectly fits the $N$ training examples. By a perfect fit we mean that the in-sample error $E_{in}(\hat{f}) = 0$. How would you expect $\hat{f}$ to perform when predicting the labels for the test set? Very likely, it will not do a very good job leading to a high out-of-sample error $E_{out}(\hat{f})$.

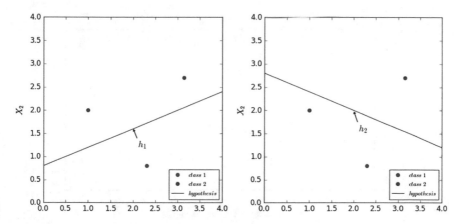

**Fig. 6.5** The points represent three input vectors (in both figures identical) and the colors their categories as separated by two example hypothesis $h_1$ and $h_2$

Thus, for a hypothesis set with an infinite VC dimension $E_{in}(\hat{f})$ is never a good estimator for $E_{out}(\hat{f})$ and therefore this hypothesis set is not PAC learnable.

The major result from the seminal work of Vapnik and Chervonenkis [122] is that all hypothesis sets with finite VC dimensions are PAC learnable and it can be shown that

$$E_{out}(\hat{f}) \leq E_{in}(\hat{f}) + \sqrt{\frac{8}{N} \log \frac{4m_{\mathcal{H}}(2N)}{N}} \qquad (6.19)$$

where $m_{\mathcal{H}}(N)$ is called the growth function that is a measure for the maximum number of elements of all possible restrictions $\mathcal{H}_X$. Equation 6.19 is called the VC generalization bound. It says, that $E_{out}(\hat{f})$ is close to $E_{in}(\hat{f})$, if and only if the term under the square root approaches zero asymptotically. With increasing $N$ the latter only happens if $m_{\mathcal{H}}(N)$ does not grow exponentially. In fact, it can be shown that whenever a hypothesis set has a finite VC dimension, the growth function can be bound by a polynomial of order $d_{VC}$. You can convince yourself that in this case the square root term in Eq. 6.19 approaches zero, the reason being that $\log m_{\mathcal{H}}(2N)$ growth slower than $N$ if $m_{\mathcal{H}}(2N)$ is polynomial.

### 6.2.3   Implications

The ideas and insights from the previous section are not only of theoretical interest but can help us when choosing an appropriate hypothesis set. Let us take a closer look at the dependency of $E_{in}(\hat{f})$ on the number of training examples. With only a few training examples we often can find $\hat{f}$ with $E_{in}(\hat{f}) = 0$. With increasing $N$ it is not possible anymore to pick $\hat{f}$ such that $E_{in}(\hat{f}) = 0$. Instead $E_{in}(\hat{f})$ will converge to a maximum (Fig. 6.6). Note, that when additional examples are used for

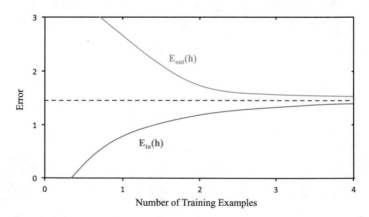

**Fig. 6.6** Dependency of in-sample and out-of-sample errors on the number of training examples

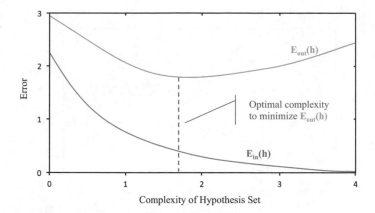

**Fig. 6.7** Dependency of in-sample and out-of-sample errors on the complexity of the hypothesis set

training purposes, $\hat{f}$ is likely to change. It is this change of $\hat{f}$ with increasing $N$ that comes with a decreasing $E_{out}(\hat{f})$. As a result and on average, $E_{out}(\hat{f})$ converges to $E_{in}(\hat{f})$—this is exactly what PAC learnability is all about.

Based on Eq. 6.19 learning, in the sense of achieving a minimum $E_{out}(\hat{f})$, can be interpreted as pursuing two objectives. First, we want to minimize $E_{in}(\hat{f})$. Second, we need to make sure that the square root term in the inequality is small. As you might have realized these are two competing objectives that have to be balanced. If we fix $N$, which it is in practice (this is, you have a limited amount of training data), we can study the dependency of in- and out-of-sample errors on the complexity of $\mathcal{H}$. While $E_{in}(\hat{f})$ decreases with increasing complexity of $\mathcal{H}$, Fig. 6.7 demonstrates that there is an optimal complexity of $\mathcal{H}$ that minimizes $E_{out}(\hat{f})$.

The mechanism behind it is that, if we choose the complexity of $\mathcal{H}$ to be too low we are not able to capture the complexity of the underlying data well. This situation is referred to as underfitting (bias of the model). On the other hand, if we choose a very complex hypothesis set, we are able to capture the complexity of the underlying data. However, we would need a lot of training data to pick the best suited hypothesis $\hat{f}$ with respect to our goal to minimize $E_{out}(\hat{f})$ (variance of the data). Since training data is limited, picking $\hat{f}$ becomes a lucky shot. This situation is called overfitting. Choosing the appropriate complexity of the hypothesis set is called Bias-Variance-Tradeoff.

The trade-off becomes obvious in Fig. 6.8. When there is only little data, the complexity of $\mathcal{H}$ (measured by its VC dimension) should be chosen to be low, because the $E_{out}^{(low)}(\hat{f})$ is expected to be smaller than $E_{out}^{(high)}(\hat{f})$. If more training data is available, it becomes beneficial to use a more complex $\mathcal{H}$ as demonstrated by the lower $E_{out}^{(high)}(\hat{f})$. As a rule of thumb the number of training examples $N$ should be at least 10 times higher than the VC dimension of the hypothesis set used for training purposes.

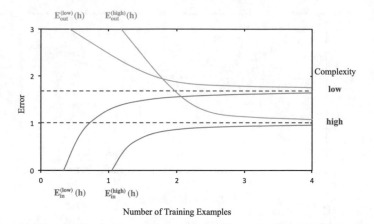

**Fig. 6.8** Dependency of in-sample and out-of-sample errors on the number of training examples for two complexity levels of the hypothesis set

## 6.3 Exercises

### 6.3.1 Pen and Paper

1. In Sect. 6.1.2 we briefly discussed the potential discrepancy between "best fitting function" and "good function". Given the discussion in Sect. 6.1.5 how would you interpret this difference?
2. In Sect. 6.1.1 we mentioned that $f$ can be thought of as the generator of observed data. Discuss this statement and the perspective on real world phenomena behind it.
3. Why is $E_{out}(h) = E(f, h)$ (see Eq. 6.6 for the definition of $E_{out}(h)$)? In which cases does the equation break?
4. Take the solution for linear regression $\hat{\theta} = (X^T X)^{-1} X^T Y$ and determine the dimensionality of the involved matrices and vectors.
5. We stated that $\nabla_\theta E_{in}(h) = \frac{X^T(X\theta - Y)}{N}$. Now it is your turn to prove this equation. (Hint: it is tricky and needs some vector calculus. A good idea is to start with one training example and then extend the number of observations)
6. Show that applying the maximum likelihood method to logistic regression reduces to the task of minimizing

$$E_{in}(h_\theta) = \sum_{j=1}^{N} (y_j - 1)\log(1 - \theta^T x_j) - y_j \log(\theta^T x_j)$$

7. Explain in detail how each value of $\Theta_{cut}$ is associated with a certain point in the ROC curve (Fig. 6.4). What if the minimum of the cost estimate lies at either $\Theta_{cut}$ 0 or 1?

## 6.3.2   Coding

1. Assume you have a linear model that generates data and you have $N < p$ observed examples. As you probably know, finding $\hat{\theta}$ is an ill-posed problem, since we cannot calculate the so-called pseudo-inverse, that is $(X^T X)^{-1} X^T$. Can we determine the in-sample error $E_{in}(h)$ anyhow? Answer the question first and then use the code provided with the book to get a graphical intuition what happens.

2. Explore gradient descent and change the learning rate $\eta$. What happens to the trajectory of the gradient descent? Is a fast gradient descent always a good idea? (Hint: the answer is No!).

3. Find a binary classification dataset and construct a ROC curve using the following Algorithm 10 which considers all unique probabilities and looks at the performance of each of these cutoff points in terms of the true- and false positive rates.

---

**Algorithm 10:** ROC curve creation

---

roc_curve_points = []
Let $P = \{< p_1, g_1 >, \ldots, < p_n, g_n >\}$ be the order set with $p_j$ the estimated probabilities for class 1 and $g_j$ the actual class
$pos = |\{< p_j, g_j > \in P | g_j = 1\}|$
$neg = N - pos$
**for** *all distinct probabilities p in P passed in descending order* **do**
  Take the set $P_{selected} = \{< p_j, g_j > \in P | p_j \geq p\}$
  Add the point $< \frac{|\{< p_j, g_j > \in P_{selected} | g_j = 0\}|}{neg}, \frac{|\{< p_j, g_j > \in P_{selected} | g_j = 1\}|}{pos} >$ to roc_curve_points
**end**
Connect roc_curve_points in order

---

Implement the algorithm.

4. Assume you have less data samples ($N$) as you have features ($p$): $p > N$. Could you still use empirical risk minimization (see Sect. 6.1.4) to explore the dependency between the input and the label space? (Hint: the answer is "yes, but ..."; besides arguing and thinking through the situation, explore the $p > N$ situation).

# Chapter 7
# Predictive Modeling without Notion of Time

After discussing various approaches to handle sensory data and to form a pre-processed dataset as well as the theoretical underpinnings of supervised learning, we can finally start making some real predictions about the quantified selves. For Arnold we would, for example, like to make predictions on his training activities in the coming week based on what we have seen in the past so that we can provide him with appropriate support. In the case of Bruce we might be very interested to fore-cast his mood during the coming week to pro-actively intervene in case of undesired predictions.

In this chapter, we will cover a variety of popular machine learning techniques that are able to generate such predictive models. The techniques discussed in this chapter do not take temporal patterns into account, but create predictive models on instances in isolation (remember that we have discussed approaches to enrich our data in order to use some of the temporal aspects of the data, even though we focus on isolated instances). We will discuss algorithms for both classification and regression. To make the book concise and self-contained, we have decided to treat the most popular algorithms on a high level. For an in-depth treatment the reader is referred to one of the many dedicated machine learning books.

## 7.1 Learning Setup

Before we explain the algorithms in detail, let us discuss the learning setup. By learning setup we refer to two aspects: (i) the split of our data into training, validation, and test data and (ii) the evaluation of the predictive performance. Remember that the training set is used to learn models, the validation set to determine when we should stop to avoid overfitting, and the test set as an independent measure of the generalizability of the model that we have created.

© Springer International Publishing AG 2018
M. Hoogendoorn and B. Funk, *Machine Learning for the Quantified Self*,
Cognitive Systems Monographs 35, https://doi.org/10.1007/978-3-319-66308-1_7

**Table 7.1** Learning setups for non-temporal learning algorithms

|  | Data | |
| --- | --- | --- |
| Level | Temporal | Non-temporal |
| Individual | $\mathbf{X_{train,qs_i}} =$ $x_{1,qs_i}, \ldots, x_{n_{train},qs_i}$ | $\mathbf{X_{train,qs_i}} \subset \mathbf{X_{qs_i}}$ where $|\mathbf{X_{train,qs_i}}| =$ $n_{train} \wedge \mathbf{X_{train,qs_i}} \cap \mathbf{X_{test,qs_i}} = \emptyset$ |
|  | $\mathbf{X_{test,qs_i}} =$ $x_{n_{train}+1,qs_i}, \ldots, x_{N_{qs_i},qs_i}$ | $\mathbf{X_{test,qs_i}} \subset \mathbf{X_{qs_i}}$ where $|\mathbf{X_{test,qs_i}}| =$ $(N_{qs_i} - n_{train}) \wedge \mathbf{X_{train,qs_i}} \cap \mathbf{X_{test,qs_i}} = \emptyset$ |
| Population - | $\mathbf{X_{train}} \subset \{\mathbf{X_{qs_1}}, \ldots, \mathbf{X_{qs_n}}\}$ where $|\mathbf{X_{train}}| = n_{train} \wedge \mathbf{X_{train}} \cap \mathbf{X_{test}} = \emptyset$ | |
| Unknown users | $\mathbf{X_{test}} \subset \{\mathbf{X_{qs_1}}, \ldots, \mathbf{X_{qs_n}}\}$ where $|\mathbf{X_{test}}| = (n - n_{train}) \wedge \mathbf{X_{train}} \cap \mathbf{X_{test}} = \emptyset$ | |
| Population - | $\mathbf{X_{train}} = \{\mathbf{X_{train,qs_1}}, \ldots, \mathbf{X_{train,qs_n}}\}$ (Note that we refer to the definition of the training sets specified in the row for the individual level here) | |
| Unseen data of known users | $\mathbf{X_{test}} = \{\mathbf{X_{test,qs_1}}, \ldots, \mathbf{X_{test,qs_n}}\}$ | |

Whether we have a temporal dataset $\mathbf{X}^{\mathcal{T}}$ or a dataset where we do not assume an ordering $\mathbf{X}$ influences our setup. In addition, as we have previously seen we might have multiple users, each contributing a dataset. We identify the datasets of the users via a subscript. We specify $\mathbf{X}^{\mathcal{T}}_{qs_1}, \ldots, \mathbf{X}^{\mathcal{T}}_{qs_n}$ for temporal datasets, and a similar notation for datasets without making any assumption about the temporal ordering: $\mathbf{X_{qs_1}}, \ldots, \mathbf{X_{qs_n}}$. Here, $qs_i$ represents user $i$. For the target we apply the same convention. For individual datapoints within a dataset, we will use $x_{i,qs_j}$ to express the $i$th datapoint of the $j$th person.

So what can we do with this data? We can generate models at two different *levels*: (i) on an *individual level*, by using only the data of a specific individual $i$ (i.e. $\mathbf{X_{qs_i}}$) or (ii) on a *population level* using data from multiple users. For the latter, we use the union of the datasets: $\mathbf{X} = \mathbf{X_{qs_1}} \cup \cdots \cup \mathbf{X_{qs_n}}$. Why would you go for a model on a population level? Well, you might have limited data available per individual, requiring a population oriented approach to have sufficient data. Obviously, if ample data is available the individual model is likely to predict in a more accurate way due to individual differences. When we consider the individual level models, the goal is clear: we want to generate a model that predicts unseen data of that individual well. On a population level we have more choices: we might be interested in predicting the target variable of completely new (unseen) users well, or we might be interested in predicting unseen data of known users accurately. Next to the different levels and purposes, we have the distinction between temporal and non-temporal datasets. If we think of the prediction of unseen data for the temporal case, we would take the time into account to split our data. For non-temporal data we can arbitrarily choose what part of the data we would like to predict. For predicting values of new users there is no difference with respect to the learning setup.

The options are depicted more formally in Table 7.1. This includes the way to split up the data in a training and test set for each one of the settings. We did not include the validation set in this table for the sake of brevity, but the way to define it is completely in line with the definition of the other two sets. We see in the table that for the individual models with temporal data we select the first $n_{train}$ data (or time) points for training and use the last $N_{qs_i} - n_{train}$ data points for testing.

For the non-temporal data on the individual level we take an arbitrary subset of our dataset of size $n_{train}$ and use the remaining points for testing. When we consider the population level models for optimizing predictive performance for new (unseen) users, we include all data from $n_{train}$ users as training data and all data of the remaining users as test set. This allows us to train on certain users and evaluate the performance of the model on new users for which we have not seen any of their data yet. For the population level case, for unseen data of known users we take the union of the training sets we have defined on an individual level over all $n$ quantified selves. This allows us to learn over all participants while we test on unseen data for each participant. Of course, the associated targets are split in the same way as our features since they are paired.

Let us move to the learning algorithms.

## 7.2 Feedforward Neural Networks

We will start our exploration with neural networks. Neural networks originated from the 1940s when Pitts and McCulloch introduced the first computational neural network. We will discuss neural networks by starting with the most simple network and will move to more sophisticated variants after that.

### 7.2.1 Perceptron

In 1958 the first learning algorithm was defined by Rosenblatt developed for a neural network model called the perceptron [100]. Figure 7.1 shows the basic layout of the perceptron, which just contains one neuron.

We can see the inputs on the left side of the figure $(X_1, \ldots, X_p)$, which are our attributes. In addition we see the so-called bias input which is constant and typically set to 1. The inputs are assumed to be numerical or binary values. The inputs and the bias are connected to the neuron via arcs that each have a weight associated with it ($w_0, \ldots, w_p$, we will express the entire vector by $w$). Based on the value of the inputs and the weights, the network provides an output $Y_1$. The computation is easy for an instance. We first take the weighted sum of the inputs:

$$v_i = w^T x_i = \sum_{n=0}^{p} w_n x_i^n \tag{7.1}$$

**Fig. 7.1** Layout of the basic
perceptron neural network

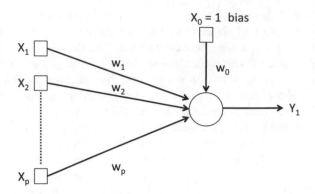

We then apply an *activation function*, $\varphi$ to determine the output of the neuron:

$$\hat{y} = \varphi(v) \tag{7.2}$$

The type of activation function $\varphi$ we use highly dependents on the target we should predict. Rosenblatt used a *signum* function (1 if $v > 0$, 0 if $v = 0$, and -1 if $v < 0$) to allow for classification. An identity function ($\varphi(v) = v$) can be used for regression, and many others exist (e.g. a sigmoid, an s-shaped function such as the *tanh*).

Note that we use $Y$ as the target variable, even though, following our notation, $Y$ refers to numerical targets. We do so because the perceptron (as most algorithms covered in this book) are designed to output numerical targets. However, a mapping from a categorical target $G$ to a numerical target $Y$ can easily be obtained. In case we have two classes, we create a single binary attribute, where 0 represents one class and 1 the other. If there are more than two classes, we can create a binary attribute per class.

Let us get back to the perceptron and use this simple network for classification. Imagine we have a dataset that contains two attributes of the data of an accelerometer (represented by the x-and y-axis). It is shown in Fig. 7.2. The red points expresses when the measurement is associated with an active user while the blue points represent an inactive one.

We want the neural network to learn to classify the data points correctly. For this we can use the training algorithm that Rosenblatt identified. The algorithm is expressed in Algorithm 11.

We assume the target is either $-1$ or 1 and process the training set in its current order. The variable between the brackets indicates the time step in the algorithm. Note that the input vector $x(n)$ also includes the bias as the first element. The parameter $\eta$ expresses the learning rate (commonly set between 0 and 1). The algorithm finds values for the weight such that the perceptron spans up a hyperplane (as shown in Fig. 7.2 in the form of a dotted line) that separates the two classes. The specification of the hyperplane is:

**Fig. 7.2** Example dataset perceptron

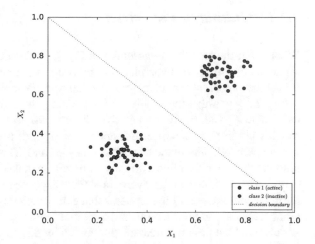

---

**Algorithm 11:** Perceptron Learning Algorithm

---

w(0) = **0**

n = 0

**while** *there are still misclassified training examples* **do**

    n = n + 1

    get the training example $x(n) = x_{n \bmod N}$ and the target value $y(n) = y_{n \bmod N}$

    compute the output of the neural network $\hat{y}(n) = signum(w(n)^T x(n))$

    adjust the weight according to $w(n + 1) = w(n) + \eta[\hat{y}(n) - y(n)]x(n)$

**end**

---

$$w^T x = 0 \tag{7.3}$$

It was shown by Novikoff [91] that given that the two classes are linearly separable by a hyperplane as described in Eq. 7.3, Algorithm 11 quickly finds a solution that fulfills $y_i = signum(w^T x_i)$ for all $i$. Do you remember the discussion from Chap. 6 about the learnability of target functions? The proof from Novikoff shows that we can reduce the in-sample error to 0 and thus approximate the unknown target function. Note that it is an approximation, since various $w$ (in fact, infinitely many) can separate the training data.

So far so good. Unfortunately, our neural network has its limits and these are severe: it can only represent linearly separable cases for classification (i.e. only classes we can separate with a hyperplane in the $p$ dimensional input space). If the classes are not linearly separable, Algorithm 11 would not terminate but jump back and forth. In case only a few training examples from the full dataset cause the trouble, there is a potential cure called the pocket algorithm that can simply stabilize learning outcome [99].

### 7.2.2   Multi-layer Perceptron

What about the situation represented in Fig. 7.3? The training examples cannot be separated by any single hyperplane (or perceptron). However, after a bit of thinking it becomes apparent that combining perceptrons $P_1$ and $P_2$ shown as dashed lines in Fig. 7.3 can solve the problem of separating the training examples. How? Let us assume $P_1$ and $P_2$ are characterized by the two vectors $w_{(1)}$ and $w_{(2)}$ which are roughly pointing to the upper left and upper right of the figure respectively (note: $w$'s are perpendicular to the hyperplane they define). Then we could classify the red points at the top of the figure correctly by requiring that $w_{(1)}^T x > 0$ and $w_{(2)}^T x > 0$. We can even generalize this: it is always (and for every potential training dataset) possible to find a combination of hyperplanes that can completely separate the training data without errors as long as there are not any input vectors in the training examples which are identical but have different labels (think why this requirement is important).

These solutions can be modeled with more complex networks such as a multi-layer neural network. We simply make multiple (usually) fully connected layers each containing one or more neurons. Connections are not present within a layer, but only between layers. These networks are called multi-layer feedforward neural networks.

We can train this type of network using the well-known back propagation algorithm. This algorithm is not necessarily guaranteed to generate the optimal solution as it can get stuck in local optima but in practice works quite well. The algorithm is based on the relationship between the error function (the mean squared error between $\hat{\mathbf{Y}}$ and $\mathbf{Y}$) and the weights. It starts with a forward pass to calculate the predicted output given the current input vectors. Errors are propagated backward in the network, thereby adjusting weights of connections that have contributed to the error more severely. It updates the weights in the direction where the mean squared error decreases.

**Fig. 7.3** Non-linearly separable case

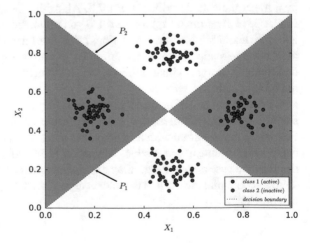

We are nearly ready to show the calculations. Assume that we have weights between neurons noted as $w_{ij}$, going from neuron $i$ in a layer to neuron $j$ in the next layer. We require an integrable activation function. Let $v_j$ be the weighed sum of the inputs of neuron $j$. Remember that we denote the calculated output of a neuron $j$ for training instance $n$ by $\hat{y}_n^j$ and the desired output by $y_n^j$. We assume that the neuron identifiers at the output match the indices of the variables. The back propagation algorithm is then expressed as follows:

$$\Delta w_{ij} = \eta \delta_j \hat{y}_n^i \tag{7.4}$$

$$\delta_j = \begin{cases} \varphi'(v_j)(y_n^j - \hat{y}_n^j) & \text{if } j \text{ is an output node} \\ \varphi'(v_j)\sum_1^k(\delta_k w_{jk}) & \text{otherwise} \end{cases} \tag{7.5}$$

In the formula, the value of $k$ is the number of neurons in the layer next to the layer where neuron $j$ is positioned. This algorithm is able to create solutions for both regression and classification problems.

### 7.2.3 Convolutional Neural Networks

Recent advances have significantly accelerated the use neuronal networks in practice. One type of neural networks that plays an important role in this development are convolutional neural networks (CNN). We will discuss the idea behind CNNs on an intuitive level. They are part of a class of networks that are often referred to as deep neural networks. They have not often been reported in the context of the quantified self so far. However, Bhattacharya et al. [15] have shown that this could be a promising avenue. So we do not want to withhold it from you.

CNNs are not very different from the multi-layer neural networks we have just discussed, except that there are layers preceding the conventional multi-layer neural network layers. These layers automatically identify features within the input space. These networks are currently mostly used for recognition of images and sound, but other applications are becoming more and more widespread. They mostly assume a 3D input space (think of the height and width of an image and the pixels R, G, and B values as a 3D input space) but 1D or 2D input spaces are also possible. For example, Bhattacharya et al. [15] use the amplitudes of the frequencies from a Fourier Transformation as an input. We will explain CNNs from a 3D input space perspective (after the explanation you will be able to see how we can tranfer this to the 1D- or 2D-space).

Two types of layers exist to identify features: convolutional layers and pooling layers. The convolutional layers contain filters that extract features from a so-called receptive field. The receptive field is a part of the input space. Each filter is represented by $k$ neurons. The receptive field is usually much smaller than the original input space. Assuming that the input is of dimension $n \times m \times r$ the receptive field is $i \times j \times q$

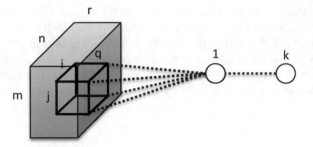

**Fig. 7.4** Convolutional layer

**Fig. 7.5** Max pooling

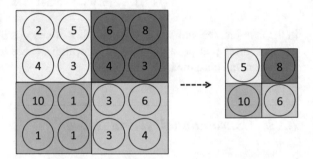

with $i$ and $j$ much smaller than $m$ and $n$ respectively, and $q$ equal or less than $r$. A filter is applied to such a receptive field (in fact, we apply the filter to each subset of the input space that matches the size of the receptive field) and gives us the values for $k$ features for that receptive field. Hence, we end up with $k$ times the number of subsets of the input space that matches the size of the receptive field. The layer is illustrated in Fig. 7.4. The filters can take different forms, for example they could extract useful features from the input space similar to what we have seen in the Principal Component Analysis earlier. These filters have to be learned, but this can be done in an unsupervised way, or we could use kernel functions which we will discuss later on. The pooling layers are simpler: they summarize the values in a certain region of the space and represent the output as a neuron. The max pooling layer is an example, which computes the maximum value of a $p \times p$ region of neurons. Figure 7.5 shows a max pooling approach for $p = 2$. After several of these layers, we find a fully connected mutli-layer neural network that can be trained using regular back propagation again.

Many more variants of neural networks exist. We will treat recurrent neural networks in the next chapter. For other variants we refer the reader to [58].

## 7.3 Support Vector Machines

An approach that uses a similar setup to the one shown for neural networks are support vector machines (SVMs) [121]. SVMs mainly target classification problems and come with very nice theoretical properties. Unlike neural networks, that aim to finding *a* hyperplane to separate classes, SVMs strive to find a hyperplane that maximizes the distance between two classes. Let us return to our previous accelerometer example to see what we mean here. Consider Fig. 7.6. The specification of the hyperplane that separates the two classes is:

$$w^T x + b = 0 \qquad (7.6)$$

This is nearly identical to the neural network based approach, except for the bias $b$ that is no longer part of the weight vector. In addition, we define two parallel hyperplanes that can be found by moving the aforementioned hyperplane towards the points of the two classes ($+1$ and $-1$ respectively) until we meet one or more data points of that class. For the class $+1$ the hyperplane specifications is $w^T x + b = 1$ for the other class it is $w^T x + b = -1$. Hence, instead of only expressing the desired value of the function for the separating hyperplane we also express the value of the function for the points positioned at the other two hyperplanes. The points that are positioned on one of these two hyperplanes are called the *support vectors*. Removing one of these points will likely impact the positioning of the hyperplane that maximizes the distance between classes. Other points will not have that effect. So how do we find this hyperplane? Well, we want to maximize the distance between the two outer hyperplanes (as shown in Fig. 7.6). This distance is equal to $\frac{2}{||w||}$ (we will not show the proof here), meaning that the magnitude of the weights should be minimized to maximize the separation between the two classes. In addition, we have the two

**Fig. 7.6** Example dataset SVMs

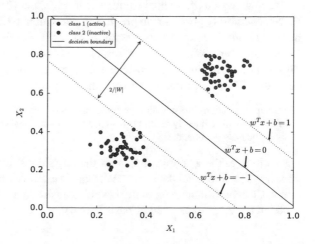

constraints on the value we have defined for the two parallel hyperplanes before (i.e. points of the class should have a value $w^T x + b \leq -1$ and $w^T x + b \geq 1$ respectively). This results in a problem that can be solved using the so-called Lagrangian multiplier. The optimization problem is expressed mathematically as follows:

$$Q(\alpha) = \sum_{i=1}^{N} \alpha_i - \frac{1}{2} \sum_{i=1}^{N} \sum_{j=1}^{N} \alpha_i \alpha_j y_i y_j x_i^T x_j \tag{7.7}$$

where

$$\sum_{i=1}^{N} \alpha_i y_i = 0 \tag{7.8}$$

$$\alpha_i \geq 0 \tag{7.9}$$

The $\alpha_i$ represent newly introduced concepts from the Lagrangian multiplier. We notice that Eq. 7.7 does not depend on the weights – they are not part of the equation. In addition, Eq. 7.7 only depends on the inner product between the instances but not the values for the attributes individually. Once we have found the maximum of the function $Q(\alpha)$ we can compute the weights as follows:

$$w = \sum_{i=1}^{N} \alpha_i y_i x_i \tag{7.10}$$

Given the weight vector $w$, finding the bias $b$ is easy: just pick a point $s$ which is a support vector for class $+1$ and use the formula for the hyperplane to compute $b$.

We now have a pretty nice solution with excellent properties. But how about the linear separability problem we previously encountered? Well, we also suffer from it here. However, for support vectors this is solved in a different way by using so-called kernel functions [39]. By using a kernel function we essentially map our inputs to a different (usually higher dimensional) feature space in which the problem is linearly separable. This is inspired by Cover's theorem:

> A complex pattern-classification problem, cast in a high-dimensional space nonlinearly is more likely to be linearly separable than in a low dimensional space, provided that the space is not densely populated.

Let us look at the example shown on the left side of Fig. 7.7 and assume that this concerns accelerometer data again. We want to classify whether the person is walking or not. Clearly we cannot separate these classes in a linear fashion. Take the following function (referred to as a radial-basis kernel):

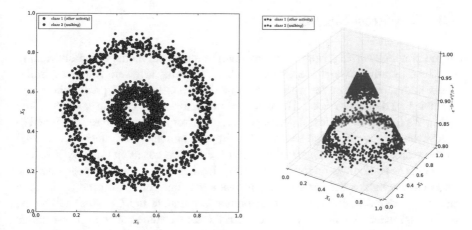

**Fig. 7.7** Example kernel function

$$K(x, x') = e^{-\frac{||x-x'||^2}{2\sigma^2}} \tag{7.11}$$

This function is a distance metric between two points. Assume that we have a fixed value for $x'$, namely the center of the circles expressed on the left side of Fig. 7.7. We calculate the kernel function (Eq. 7.11) of each data point $x$ with respect to $x'$. The results are stored in a new attribute called $X_3$ displayed on the right side of Fig. 7.7. We can now fit a hyperplane that separates the classes by means of the newly introduced values: lower values for $X_3$ are other activities while higher values indicate walking. This is caused by the points represented in blue (walking) all lie closer to the center (our center point $x'$) compared to all points in the other circle (other activity). The advantage of support vector machines is that we do not have to calculate all the points in the new feature space after application of the kernel function, but we can perform calculations in the initial feature space. This is referred to as the *kernel trick*:

$$Q(\alpha) = \sum_{i=1}^{N} \alpha_i - \frac{1}{2} \sum_{i=1}^{N} \sum_{j=1}^{N} \alpha_i \alpha_j y_i y_j k(x_i, x_j) \tag{7.12}$$

where $k$ is a kernel function. Many options for kernel functions are available (such as the one we have seen before). SVMs aim to solve a binary classification problem. Extensions have also been proposed that allow regression problems to be represented by SVMs as well, called Support Vector Regression (SVR) [108].

## 7.4   K-Nearest Neighbor

In the previous two algorithms, we have built a statistical model based on training data by estimating a weight vector $w$. Another possibility is to simply look at similar cases in the input space for which we know the target value and predict the target variable based on these cases. The k-nearest neighbor approach is an example of the latter. We look at the $k$ closest points $[x_1, \ldots, x_k]$ from the training set compared to a new example $x$ and assign a class value $\hat{g}$ based on the target values $[g_1, \ldots, g_k]$ of those neighbors (or $\hat{y}$ in case of regression problems based on $[y_1, \ldots, y_k]$). The algorithm is sometimes referred to as a lazy learner, and essentially postpones the computation until a new case comes in. As a starting point, we need to define a metric to compute distances between points to compute the $k$ nearest neighbors, and an aggregation function to come from the target values of those $k$ neighbors $[g_1, \ldots, g_k]$ to $\hat{g}$ or from $[y_1, \ldots, y_k]$ to $\hat{y}$. We have extensively looked at distance functions when we considered the clustering in Chap. 5 and will not treat them in detail again. It suffices to say that we have some distance function $d(x_i, x_j)$ between two points $x_i$ and $x_j$.

Let us focus on classification first. As our prediction we assign the majority class. For the math enthusiasts, we define the majority class as follows:

$$\hat{g} = \underset{v \in \mathcal{G}}{\operatorname{argmax}} \sum_{i=1}^{k} \delta(v, g_i) \tag{7.13}$$

where

$$\delta(v, g_i) = \begin{cases} 1 & \text{if } v = g_i \\ 0 & \text{otherwise} \end{cases} \tag{7.14}$$

When there is a tie we can arbitrarily select a class. In Fig. 7.8 we consider our previous example and a new point we have to classify (the black dot). If we select $k = 3$ neighbors the points indicated by the yellow star are the 3 nearest neighbors given a Euclidean distance function. Since the majority is blue (i.e. the *inactive* class), this is the assigned class.

For a regression problem, the computation is even more straightforward. We simply take the average of the values $[y_1, \ldots, y_k]$ of the $k$ nearest neighbors:

$$\hat{y} = \frac{\sum_{i=1}^{k} y_i}{k} \tag{7.15}$$

One aspect we have not taken into account is the distance of a neighbor to the point we want to determine the value for. Closer points might be considered more important as they are more similar. This is the basic premise of distance weighted

**Fig. 7.8** Example k-nearest neighbors

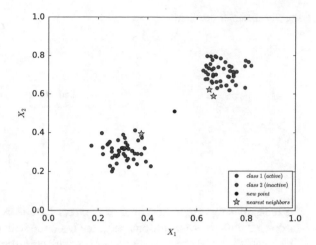

nearest neighbor. We can redefine the equations we have seen for classification and regression by assigning a weight $w_i$ to each case:

$$\hat{g} = \underset{v \in \mathcal{G}}{\operatorname{argmax}} \sum_{i=1}^{k} w_i \delta(v, g_i) \qquad (7.16)$$

$$\hat{y} = \frac{\sum_{i=1}^{k} w_i y_i}{\sum_{i=1}^{k} w_i} \qquad (7.17)$$

where the weight is defined as:

$$w_i = \frac{1}{d(x, x_i)^2} \qquad (7.18)$$

Numerous variants of the algorithm have been introduced, the reader is referred to [17] for a more extensive overview.

## 7.5 Decision Trees

Thus far the attributes that we use as predictors have mostly been numerical or binary attributes, although k-nearest neighbor can obviously handle any type of attribute depending on the distance function used. Decision trees are a prime example of algorithms that handle all types of attributes in a very natural way. Figure 7.9 shows an example of a decision tree with attributes taken from the dataset produced by Arnold. In the figure, the oval shapes represent the *nodes* of the decision tree, while

**Fig. 7.9** Example decision
tree

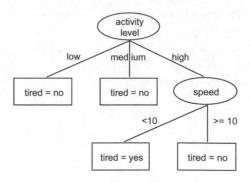

the rectangles represent the *leaves* of the tree. A node in the tree is a decision point
associated with a certain attribute (e.g. the top node with the attribute *activity level*).
The branches (lines) show values associated with the attribute (e.g. *low*, *medium*, and
*high*). For numerical attributes the values in the branches represent a condition (e.g.
< 10). The leaves assign a value for a class or a numerical value for cases that have
fulfilled all conditions from the top node to the leave. A case with a value for *activity
level* = *high* and *speed* = 9 would for instance be classified as *tired* = *yes*.

Obviously, we strive to find a tree that is able to present our data in a generalizable
way. We follow the ID3 algorithm [96]. Note that many other decision tree learning
algorithms exist [57]. We will start with an empty tree and build the tree by starting
with the most important attribute. But how do we define this importance? We will see
this soon. Here we assume that we have a measure of importance for an attribute $X_i$
in our dataset. The algorithm for building a tree for a classification task is specified
in Algorithm 12.

---

**Algorithm 12:** Decision Tree Learning Algorithm

---

   **build_tree(X, G):**
   **if** *stop condition reached* **then**
     |   return a leaf node the majority class in $G$
   **end**
   $X_{best} = \text{argmax}_{X_i \in X}$ importance$(X_i, \mathbf{X})$
   create a node for $X_{best}$
   **for** *all values v in the domain of $X_{best}$* **do**
     |   create a branch from the create node labeled with v
     |   set the tree under the branch with build_tree($\{x_i \in \mathbf{X} | x_i^{best} = v\}, \{y_i \in \mathbf{G} | x_i^{best} = v\}$)
   **end**
   **return** *the created node*

---

Essentially, the algorithm creates branches recursively until a stop condition has
been reached (e.g. minimal number of cases in $\mathbf{X}$). It creates a node that represents
the most important attribute in the part of the training set and creates branches for
each possible value of the attribute. For each branch a new tree is formed based on
the subset of the training set that contains the associated attribute-value combination.

We still have a number of unanswered questions, the first being in what way we can process numerical attributes using this algorithm. The various decision tree algorithms handle this problem in different ways, but we will consider a discretization procedure following ID3 [96]. Using this procedure we can transform our numerical attributes to categorical attributes with values that represent interval points of our numerical attribute (Algorithm 13).

---

**Algorithm 13:** Discretization procedure

Let $< x_1^i, g_1 >, \ldots, < x_N^i, g_N >$ be the continuous values for an attribute $X_i$
    observed in $N$ training instances paired with their class values
Sort the pairs low to high based on the value for $X_i$
    respectively in the sorted list
interval_points = []
previous_class = $g_1$
**for** $j$ in $2, \ldots, N$ **do**
    **if** $g_j$ *is not equal to previous_class* **then**
        Add $\frac{x_{j-1}^i + x_j^i}{2}$ to interval_points
        Set previous_class to $g_j$
    **end**
**end**

---

The algorithm sorts the values of the continuous attribute and creates interval points at values where the target changes. Table 7.2 illustrates the procedure for two attributes *speed* and *heart rate* given the target *tired*. Here, interval points are created at *speed* = 5 and *speed* = 9.5. So which interval point should we select? Since we want to select a point that creates the best tree, we try a discretization using each of the interval points and determine which one provides most information.

We have been referring to the best attribute and most information a lot. How do we define this? Here, a common approach is to use the *information gain*. This measure is based on the notion of *entropy* introduced by Shannon [106]. The entropy is related to the amount of bits that are required to send a certain message. The more information the message contains, the more bits are required. Consider our classification problem. Let us say that we want to send information about the class of a series of instances. If all instances are of the same class this will result in minimal information, i.e. an entropy of 0. If we have a binary classification problem and our instances are evenly spread over the two classes we have the maximum amount of information. We need to communicate half of our instances in order to classify

**Table 7.2** Discretization example

| Speed | 0 | 0 | 0 | 5 | 5 | 9 | 10 | 10 |
|---|---|---|---|---|---|---|---|---|
| Tired | no | no | no | no | yes | yes | no | no |
| Interval points Speed | – | | | – | | | – | |

them correctly, it can never be more. The entropy will be 1. Assume that we have probabilities $p_1, \ldots, p_n$ associated with each of $n$ classes, representing the relative occurrence of each class in the data. The entropy is then defined as follows:

$$entropy(p_1, \ldots, p_n) = \sum_{i=1}^{n} -p_i log_2 p_i \qquad (7.19)$$

We can define the entropy of a dataset with targets $\mathbf{G}$ as:

$$entropy(\mathbf{G}) = \sum_{v \in \mathcal{G}} -p_v log_2 p_v \text{ where } p_v = \frac{|\{g_i \in \mathbf{G}|g_i = v\}|}{|\mathbf{G}|} \qquad (7.20)$$

Given the knowledge we have built up in this section, we know that we should strive for: leaves that cover a set of instances of the same class, i.e. an entropy of 0. We start with our whole training set, which is likely to have a high entropy. Step by step we should select attributes for our nodes such that we get closer to our goal, i.e. we should split our data in such a way that we end up with splits of our training set that have a low entropy. Hence, we should select attributes that reduce the entropy as much as possible. Based on these considerations we define define the *information gain*:

$$gain(X_i, \mathbf{X}, \mathbf{G}) = entropy(\mathbf{G}) - \sum_{v \in \mathcal{X}_i} \frac{|\{x_j \in \mathbf{X}|x_j^i = v\}|}{|\mathbf{X}|} \cdot entropy(\{g_i \in \mathbf{G}|x_j^i = v\}) \quad (7.21)$$

In the formula, we look at the original entropy of the set with labels $\mathbf{G}$ and the new entropy if we create subsets by means of the values of the attribute $X_i$. The entropy for each subset is calculated and weighed according to the size of the subset. Note that ample improvements have been made to it [97].

We end this section by looking at decision trees that can be used for regression. To do so, we need to change the criterion for splitting nodes and define what value we assign to new cases in the leaves of the tree. Let us consider the value assignment in the leaves first. Two variants exist. *Regression trees* assign the average value of the instances of the training set that fulfil the criteria of a leaf. The other variant, called *model trees* perform a regression on the relevant instances and use the regression equation to assign a value to new cases.

Moving on to the splitting criterion we see that our information gain measure does not work any more, since it requires categorical targets. An alternative for regression is the *standard deviation reduction*, defined as:

$$sd\_red(X_i, \mathbf{X}, \mathbf{Y}) = \sigma_Y - \sum_{v \in \mathcal{X}_i} \frac{|\{y_i \in \mathbf{Y}|x_j^i = v\}|}{|\mathbf{Y}|} \sigma_{\{y_i \in \mathbf{Y}|x_j^i = v\}} \qquad (7.22)$$

The underlying principle is that attributes that reduce the standard deviation (i.e. the diversity in the target values) are preferred as instances will have more similarity in terms of the target value. A similar weighting as introduced for the information gain (based on the size of the subsets) is applied.

## 7.6  Naive Bayes

The final basic method we will discuss in this chapter is Naive Bayes. Naive Bayes is targeted towards classification problems. Classification is based on conditional probabilities of the form $p(g|x)$, also called *posterior probability*. Remember that $x$ represents an instance that we want to classify and $g$ the corresponding class (i.e. $g \in \mathcal{G}$). $p(g|x)$ expresses the probability of observing class $g$ given the observation $x$. How can we derive these conditional probabilities? Let us start with introducing some calculations based on the data we have observed in our training set. We can calculate $p(g)$, the *prior probability*, i.e. the probability of observing a class $g$ within the dataset, as follows:

$$p(G = g) = \frac{|\{g_i \in \mathbf{G}|g_i = g\}|}{N} \qquad (7.23)$$

This is simply the number of observations of class $g$ divided by the total number of observations. We can calculate the probability that we observe $x$ given a class $g$ (i.e. $p(x|g)$) as follows:

$$p(x|G = g) = \prod_{i=1}^{p} p(X_i = x^i|G = g) \qquad (7.24)$$

where

$$p(X_i = v|G = g) = \frac{|\{j \in 1, \ldots, N|(x_j^i = v \wedge g_j = g)\}|}{|\{i \in 1, \ldots, N|g_i = g\}|} \qquad (7.25)$$

This is called the *class conditional probability*. In Formula (7.24) we encounter the naive assumption in Naive Bayes (hence the name): we multiply the conditional probabilities of individual attributes. We can do this under the assumption that the attributes are conditionally independent. Obviously, this will not always hold: there might be correlations between the attributes. However, it does simplify matters a lot and in practice works quite well. We also see the consequences of this assumption when we calculate the probability of observing the values for $x$ (also called the *evidence*):

$$p(x) = \prod_{i=1}^{p} p(X_i = x^i) \qquad (7.26)$$

where

$$p(X_i = v) = \frac{|\{j \in 1, \ldots, N | x_j^i = v\}|}{N} \tag{7.27}$$

We are now ready to calculate the posterior probability according to Bayes rule:

$$p(G = g | x) = \frac{p(x | G = g) p(G = g)}{p(x)} \tag{7.28}$$

The class we assign is the one with the highest probability:

$$\hat{g} = \underset{g \in \mathcal{G}}{\operatorname{argmax}} \, p(G = g | x) \tag{7.29}$$

A disadvantage of the Naive Bayes approach is that the assigned probability of a target drops to zero when we have an attribute-value pair we have not seen in combination with the target. This might be a bit too harsh as it could for example be caused by a lack of sufficient training data. The Laplace estimator can help here. We simply add 1 to the numerator and the number of values of the attribute to the denominator, i.e.

$$p(X_i = v | G = g) = \frac{1 + |\{j \in 1, \ldots, N | (x_j^i = v \wedge g_i = g)\}|}{|\mathcal{X}_i| + |\{j \in 1, \ldots, N | g_j = g\}|} \tag{7.30}$$

This gives a non-zero probability in case we have no observations and approaches the actual probability when $N$ becomes large. If we face a missing value for a certain attribute we can simply ignore that attribute in the calculations, a very natural and simple way of handling missing values. This is one of the advantages of the Naive Bayes classifier.

So far we have only considered categorical attributes. Naive Bayes is able to cope with numerical values as well, and in fact it does so in a very natural way. Instead of using a probability mass function for discrete attributes, we describe the data in terms of a probability density function (*pdf*) to represent $p(X_i = v | G = g)$, e.g. the normal distribution. When evaluating $p(X_i = v | G = g)$ for an unseen data point the *pdf* returns a numerical value reflecting the probability to observe this value. Thus, the result can be used to predict the most likely class with Eq. 7.29.

## 7.7   Ensembles

It is time to kick it up a notch. We have seen a variety of relatively simple models. But simplicity comes at a price. We might not be able to achieve a very good predictive performance. There are two main causes for that: (1) the expressive power of the models might be insufficient causing a hampered performance (bias of the model, see Sect. 6.2.3), or (2) the training data is simply too limited (variance of the data). Ensembles can help out here. The idea behind ensemble learning is not to create a single model using the algorithms we have considered but a number of these models. In the end, the ensemble of models returns a combined answer that is the result of a voting scheme for classification tasks or averaging for regression problems. Two main streams of ensemble methods exist: bagging approaches (aimed at reducing variance) and boosting approaches (you probably guessed it, aimed at reducing the bias).

### 7.7.1 Bagging

In general, bagging (or bootstrap aggregation [29]) approaches are easy to understand. They draw $m$ samples $\mathbf{X}_i$ of size $n$ (with replacement) from the original dataset $\mathbf{X}$. For each of the samples $\{\mathbf{X}_1, \ldots, \mathbf{X}_m\}$ we generate a model: $\{M_1, \ldots, M_m\}$ (Algorithm 14).

---

**Algorithm 14:** Bagging

---

models = []
**for** *i* in *1,...,m* **do**
  Take a sample $\mathbf{X}_i \subset \mathbf{X}$
  Build a model $M_i$ on training data $\mathbf{X}_i$ using some learning algorithm
  Add $M_i$ to models
**end**

---

In the end, we combine the output using majority voting as we have seen for k-nearest neighbor. The sampling avoids overfitting of the models towards the data. A prime example of a bagging approach using decision trees is the *random forest* technique. The random forest technique generates $m$ trees based on the aforementioned samples. In addition, it uses a selection of the attributes when creating the decision trees: when an attribute needs to be selected in the process of building a tree only a subset of the available attributes is considered. This reduces the chances that all trees strongly resemble each other (e.g. if there are a few dominant predictors) since very similar trees combined are not likely to predict a lot better than a single tree.

### 7.7.2 Boosting

As said before, boosting aims to tackle the problem of bias. To do this, it iteratively creates models that focus on the areas where mistakes are being made by previously generated models. Take a look at Fig. 7.10.

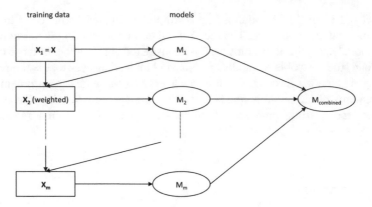

**Fig. 7.10**  Boosting process

In the figure, we see that we start with our initial training set **X** and build an initial model $M_1$ on it using some learning algorithm. Then, we look at the performance of $M_1$ on the training set, and form a new training set **X₂** which weights cases where we made mistakes with our previous model more heavily. This process is repeated $m$ times. We get specialized models that focus on subsets of the data. Together these models have more expressive power than a single model, meaning that a single model might not be able to grasp all cases in the training set equally well. After the process has been completed, we end up with $m$ models (i.e. $\{M_1, \ldots, M_m\}$) that form an ensemble and combine their output in the same way as we have discussed before.

Let us take a closer look into how we can compose new training sets. We consider an example of a boosting approach aimed at classification first. It is called Arcing [49]. Arcing works on the principle of assigning sampling probabilities $\{p_1, \ldots, p_N\}$ to each of the $N$ data points of the training set. As said, we build $m$ models (i.e. take $m$ steps), and in the $k$th step the probabilities (with the identification of the step in brackets, e.g. $p_1(k)$) are set based upon the error of the ensemble we have built so far. Let $d_i$ denote whether the training example $x_i$ is *incorrectly* classified with the current model $k$ ($d_i = 1$) or not ($d_i = 0$). We define the weighted misclassification rate $\epsilon$ in step $k$ as follows:

$$\epsilon_k = \sum_{i=1}^{N} p_i(k)d_i \tag{7.31}$$

The factor $\beta_k$ is defined to compute the new weights:

$$\beta_k = \frac{1 - \epsilon_k}{\epsilon_k} \tag{7.32}$$

The higher the weighted misclassification rate $\epsilon_k$ the lower $\beta_k$. The algorithm assumes $\epsilon_k \leq \frac{1}{2}$ otherwise it terminates. The termination criterion essentially says that the execution stops when you are predicting with a higher misclassification rate than a random choice for a balanced binary classification problem. This means that $\beta_k$ resides in the interval $[1, \infty]$ where 1 means the highest misclassification rate (i.e. $\frac{1}{2}$) going to $\infty$ for very low weighted misclassification rates. The probabilities are updated as follows:

$$p_n(k + 1) = \frac{p_n(k)\beta_k^{d_n}}{\sum_{i=1}^{N} p_i(k)\beta_k^{d_i}} \tag{7.33}$$

This shows that the sampling probability of a training instance $n$ in the numerator remains the same for correctly classified cases (i.e. $\beta_k^0 = 1$) and given that $\beta_k \geq 1$ will result in a higher sampling probability for incorrectly classified cases. These probabilities increase when $\beta_k$ becomes higher (i.e. when there is a very low misclassification rate). The intuition is that for those cases there are only limited instances that have been misclassified, so their weight needs to be severely increased compared to the correctly classified cases. Finally, we assign a weight to each classifier $k$ in the voting proportional to the value of $\beta_k$, the weight being $log(\beta_k)$. We then end up with Algorithm 15.

---

**Algorithm 15:** Arc-fs

models = []
probabilities = []
Set $p_1(1), \ldots, p_N(1) = \frac{1}{N}$ and add them to probabilities
**for** $k$ in $1,...,m$ **do**
    Create $\mathbf{X_k}$ by sampling from $\mathbf{X}$ with probabilities $p_1(k), \ldots, p_N(k)$ for training instances
    Create model $M_k$ using $\mathbf{X_k}$ as training data
    Determine the misclassification of $M_k$ $(d_n)$ for all training instances $n = 1, \ldots, N$
    Calculate $\epsilon_k$ using Eq. 7.31
    Calculate $\beta_k$ using Eq. 7.32
    Calculate the probabilities $p_1(k + 1), \ldots, p_N(k + 1)$ using Eq. 7.33 and add them to probabilities
    Add $M_k$ to *models* with $log(\beta_k)$ as weight
**end**

---

When we apply boosting to regression, life is a bit easier. We can simply consider the difference between the predictions and the actual values, and train a model on those differences. This is expressed in Algorithm 16. To predict the output for a new case $x$, we return the sum of the models applied to the case.

---

**Algorithm 16:** Boosting for regression problems

---

models = []
$Y_1 = Y$
**for** $k$ *in* $1,...,m$ **do**
    Build a regression model $M_i$ based on $X$ and $Y_k$
    Add $M_i$ to models
    **for** $i \in 1, \ldots, N$ **do**
        Compute the prediction of $M_i$ for $x_i$: $\hat{y}_i$
        Add the difference $y_i - \hat{y}_i$ as new desired value to $Y_{k+1}$
    **end**
**end**

---

## 7.8 Predictive Modeling for Data Streams

The algorithms described earlier in this chapter are based on the same assumptions discussed in Chap. 5 (unlimited storage, no conceptual drift). Let us consider a classification or regression task for which the assumptions listed above are not realistic. Two different solutions are suggested within data stream mining (cf. [1], pp. 43): (1) *data-based solutions*, building models on a subset of the full dataset, and (2) *task-based solutions* that focus on changing the algorithms to make them more efficient. Examples of data-based solutions are sampling and aggregation while approximation algorithms (less precise as the algorithms we have seen before) is an example of the task-based solutions. Let us look at two example algorithms, one for each solution type.

The first algorithm (an example of a data-based solution) is based on the ensemble learning approach we have just considered [123]. It assumes that the data can be split into a number of chunks. Assuming $n$ chunks, we refer to them as $S_1, \ldots, S_n$, $S_n$ being the most recent data. Chunks contain a certain number of instances that have arrived in sequence (remember that we assume an ordering of instances in data streams). For a windows size of $\omega$, $S_1$ would contain the first $\omega$ elements that arrive: $x_1, \ldots, x_\omega$, and $S_i$ contains $x_{i-1 \cdot \omega+1}, \ldots, x_{i \cdot \omega}$. For each of the chunks we build a model and do not need to store data beyond our window size $\omega$. We just save the models and refer to them to as $M_1, \ldots, M_n$. The weight of the models in the ensemble is set proportional to the error it makes on the most recent data $S_n$: the higher the error on the most recent data, the lower the weights of the classifier in the ensemble. In this way, we do not ignore our old models, but do reduce their influence if they become obsolete.

The second algorithm, which exemplifies the task-based solutions, concerns a decision tree learner called *Hoeffding Trees* [41]. The algorithm assumes that examples can only be seen once and data comes in at a continuous rate. If we want to build our tree, we need to decide on the most important attribute we will use as the root of the tree. We will use the first instances of our stream for this. Once we have decided, we will use the succeeding instances to decide on how to proceed further down the tree. So when are we certain enough about the choice for an attribute? When do we have enough data to decide? For this we use the Hoeffding bound (see Chap. 6.2.1).

We want to have a measure that provides us a real-value $r$ on the suitability of an attribute with range $R$ (for a probability this would be $[0, 1]$, for the information gain $log(|\mathcal{G}|)$ where $|\mathcal{G}|$ represents the number of classes). Let $n$ be the number of instances we have seen so far. We want to say with probability $1 - \delta$ that the mean of the variable $r$ is within distance $\epsilon$ from the true mean. We define Hoeffding bound $\epsilon$ by rearranging Hoeffding's inequality from Eq. 6.15

$$\epsilon = \sqrt{\frac{R^2 ln(\frac{1}{\delta})}{2N}} \tag{7.34}$$

If we consider the distance of our best and second best attribute computed based on $N$ instances (e.g. the difference between the highest and second highest information gain calculated), we can create a node in our tree when the difference between the two scores is greater than the computed Hoeffding bound $\epsilon$. We then consider the new instances (we can throw away the $n$ instances we used before) and continue building our tree, and end up with a nice tree for which there was no need to store all data. This approach is based on the assumption that the data is *stationary*, i.e. we do not assume *temporal locality*, unlike the previous approach we explained.

The final option of data stream mining we want to discuss is reducing the amount of data by making a selection of relevant instances that we can use to build our model at certain time points. As said before, we can no longer assume that we can store everything. So which instances should we keep then? Many strategies exist. Usually, a fixed window is used. We call this a *sliding window* since we only store the $n$ most recent instances, continuously replacing the oldest instance with the incoming one (or only do that after every ith instance that arrives to "slow down" the data). Other approaches replace instances that are stored in the window with new instances with a certain probability.

## 7.9   Practical Considerations

We have seen a lot of different approaches, but will they just work without any effort by simply applying them to the pre-processed data? Well, not quite. This section will discuss some practical considerations concerning the question how we can get the model that generalizes best. We will discuss feature selection and regularization for this purpose.

### 7.9.1   Feature Selection

While some approaches are able to cope with large sets of features (e.g. a random forest), the performance of other approaches (e.g. a decision tree) can severely degrade when features are present that have limited predictive power, they tend to result in overfitting on the training data. Therefore, we want to select those features that come with predictive power.

A common approach is to compute the *Pearson coefficient* to determine the correlation of an attribute with the target. We select the $n$ attributes with the highest magnitude (which could be a positive or negative correlation of course). We compute the value in the following way:

$$\rho(X_i) = \frac{\sum_{j=1}^{N}(x_j^i - \bar{X}_i)(y_j - \bar{Y})}{\sqrt{\sum_{j=1}^{N}(x_j^i - \bar{X}_i)^2}} \sqrt{\sum_{j=1}^{N}(y_j - \bar{Y})^2} \tag{7.35}$$

The numerator of the equation expresses the covariance between the attribute $X_i$ and the target $Y$ which is assumed to be a single numerical value. $\bar{X}_i$ and $\bar{Y}$ denote the mean. The denominator is the product of the variance of the attribute and the target. $\rho(X_i) = 1$ expresses the maximum positive correlation, -1 the maximum negative correlation, and 0 no correlation. Using the Pearson coefficient is very simple but could ignore more complex dependencies, e.g. multiple features, that only together exhibit some predictive power.

---

**Algorithm 17:** Forward selection

---

```
selected_attributes = {}
performances = []
for k = 1, ..., p do
    best_attribute = ""
    best_performance = ∞
    available_attributes = X \ selected_attributes
    for l ∈ 1, ..., |available_attributes| do
        temp_attributes = selected_attributes ∪ available_attributes_l
        performance = learn_model(temp_attributes, X)
        if performance < best_performance then
            best_attribute = available_attributes_l
            best_performance = performance
        end
    end
    performances[k] = best_performance
    selected_attributes = selected_attributes ∪ best_attribute
end
return performances
```

---

An alternative approach is an iterative process to compose a suitable set of attributes. Let us first focus on *forward selection* (for instance used by [74] to select features for context recognition). We start with an empty set of attributes and iteratively add the attribute that results in the best performance. Here, performance can be measured on the training set. Assuming we obtain a performance score when we provide the learning algorithm with a set of attributes and a dataset (a low value of the performance score is assumed to be good, e.g. the mean squared error), our selection method can be written according to Algorithm 17. This returns performance scores under varying numbers of attributes. For computational reasons, we select one attribute at a time and do not consider all possible combinations (it is a heuristic). In the case study, we will see an example of the influence of the number of selected attributes upon the performance.

Besides forward selection, we can also use *backward selection*. This works according to the same principles except that we start with the set of all attributes and iteratively remove attributes that have the least impact on performance. See Algorithm 18.

---

**Algorithm 18:** Backward selection

selected_attributes = $X$
performances = []
performances[p] = learn_model(selected_attributes, $\mathbf{X}$)
**for** $k = p - 1, \ldots, 1$ **do**
    worst_attribute = ""
    best_performance = $\infty$
    available_attributes = selected_attributes
    **for** $l \in 1, \ldots, |available\_attributes|$ **do**
        temp_attributes = selected_attributes \ available_attributes$_l$
        performance = learn_model($temp\_attributes$, $\mathbf{X}$)
        **if** $performance < best\_performance$ **then**
            worst_attribute = available_attributes$_l$
            best_performance = performance
        **end**
    **end**
    performances[k] = best_performance
    selected_attributes = selected_attributes \ worst_attribute
**end**
**return** *performances*

---

## 7.9.2 Regularization

As we have discussed in Chap. 6 we want to learn models that are generalizable, i.e. that allow us to make accurate predictions for unseen data, not just for the training data. We have discussed that more complex models tend to overfit. We could try to overcome this problem by selecting attributes based on a performance score as we have seen previously.

A widely used alternative in machine learning is regularization. The basic assumption behind regularization is that less complex models and small model parameters should be preferred (Occam's razor). To apply regularization, we add a regularization term to the objective function (such as the one in Eq. 6.5) of our learning algorithms. This term punishes the complexity of the model and allows us to strike a balance between in-sample error and model complexity during the training process. In case of a linear regression (Eq. 6.9) the objective function would be modified as follows:

$$E_{in}(h) = \frac{1}{N}(\mathbf{Y} - \theta^T \mathbf{X})^T (\mathbf{Y} - \theta^T \mathbf{X}) + \Lambda(\theta) \qquad (7.36)$$

The term $\Lambda(\theta)$ that we have added is called the regularizer. Several choices for $\Lambda(\theta)$ are possible. In practice, two specific regularizer are often used. First, $\Lambda(\theta) = \lambda\theta^T\theta$ where $\lambda$ is a real-value constant. The higher the value for $\lambda$ the less complex will the model be. Why is that? As $\lambda$ increases the importance of regularizer for the objective function in Eq. 7.36 increases. Therefore higher values for the model parameters $\theta$ are punished even if they would have minimized the original objective function (Eq. 6.9). This regularized version of the linear regression is called ridge regression (or Tikhonov regularization).

Second, $\Lambda(\theta) = \lambda||\theta||_1$ where $||\theta||_1$ is the so-called $L_1$ norm that simply adds the absolute values of the elements of the vector $\theta$. Compared to the first regularizer we have discussed, there is a striking difference that is far from obvious: While $\lambda\theta^T\theta$ tries to keep model parameters small, the second regularizer $\lambda||\theta||_1$ shrinks the influence of some of the features to 0. Thus, the higher the value for $\lambda$ the less features will be included in the model, a convenient mechanism for selecting features. This regularized version of the linear regression is called LASSO (least absolute shrinkage and selection operator) regression [117]. Recently, both regularizers have been used jointly under the term elastic net regularization [133]. One final question remains: How do we select the best $\lambda$? In practice, $\lambda$ is systematically explored (varying it from 0 to a large positive value). For each value $\lambda$ we apply cross validation to determine the predictive performance of the corresponding model and finally choose the $\lambda$ with the best performance.

We hope you are as excited about regularization as we are, since it is a basic ingredient in many machine learning algorithms. So we recommend to study it in more detail, if you want to dive deeper into the field, see e.g. [57].

## 7.10 Case Study

For the crowdsignals dataset we have identified two tasks in Chap. 2: a classification task, namely the prediction of the label representing the activity of a user, and a regression task revolving around the prediction of the heart rate. We will study the performance of the algorithms we have introduced in this chapter upon these two tasks.

## 7.10.1 Classification: Predicting the Activity Label

The first task we address is the prediction of the label associated with an instance. Since we only have data of a single user, the prediction resides on the individual level. We are going to consider this task by looking at the data instances in isolation, i.e. we will not consider the order of the instances explicitly, although we will use the temporal features we have identified in Chap. 4. We will therefore sample instances for our training and test set randomly and do not take the first period of data for training and the remainder of the data for testing. We made this choice because we have limited data available. We mostly see single (connected) periods in which a certain label has been assigned. Hence, if we would split based on time we would train on labels that are not part of the test set and we would test on labels we did not train upon. This would be a violation of the requirements in Sect. 6.1, namely, that the joint distribution $p(x, y)$, and especially $p(x)$, of the training set should be identical or at least similar to the one of the test set. In addition, it is interesting to see how well we are able to detect activities solely based on the sensory data at a certain time point, or just a few time points before. As a performance metric we use the accuracy.

### 7.10.1.1    Preparing the Dataset for Learning

So how do we go about this task? Well, first of all we need to define the target of the classification task. While we previously used a combination of binary attributes to represent the label (i.e. an attribute for each label value) we will now merge these into a single categorical attribute. This merging is not always trivial. There is a trivial case though: if we have a value of 1 for one label attribute in the instance and a 0 for all others we can safely assign that label. However, it might be that more than one label has been assigned to an instance, or no label at all. In such cases, we will assign an 'unknown' label. Once we have assigned the label, we remove the cases with the 'unknown' label: we do not want to pollute our dataset with labels we do not know the correct value of. An alternative could be to look within the intervals to see what activity was indicated for most of the time. Given the level of granularity and level of precision this does not sound very reasonable. In addition, if we would keep the label 'unknown', the learning process could be disturbed, since some of the cases might actually be clear examples for other labels. Hence, the algorithm would not be able to learn those labels properly anymore.

We end up with a cleaned set of attributes and accompanying classes with a total of 1837 out of the previous 2895 instances. Since we have sufficient instances there is no need for a cross validation scheme: we split the dataset into a training set (70% of the data, 1285 instances) and a test set (the remaining part, 552 instances). In order to learn on a representative set, we do so in a *stratified* way, making sure that the labels are represented equally frequent in both the training and the test set.

### 7.10.1.2 Selecting Attributes

Next, we select the attributes we are going to use to predict. We use different subsets of attributes to study the their added value. The subsets are shown in Table 7.3. The basic features contain a total of 21 features that we initially described in Chap. 2. These are the cleaned variants (i.e. outliers removed, missing values imputed, and the lowpass filter applied to the period attributes). In addition, we have the PCA features where we use 7 components that explain nearly all variance. We created time-based features in Chap. 4 using the mean and standard deviation over a short historic window for all the 28 features we identified before, resulting in $28 \times 2 = 56$ additional features. From the frequency domain we selected 18 measurements that are likely to exhibit periodic patterns. For each of these features we apply 3 aggregation functions for the frequency domain and have the amplitudes of 21 different frequencies. An observant reader might spot that we use a windows size of 40, so we would expect to see 41 different frequencies. The package we use however limits the output to half of the frequencies. This has to do with the real and imaginary part of the periodic functions. Discussing this is beyond the scope of this book. This totals to $18 \times 24 = 432$ features. Finally, we add a feature representing the cluster the data instance is part of (cf. Chap. 5). We do not use the temporal patterns that result from the labels as this would be cheating: it contains information about the current label, and hence the class we aim to predict, as well.

For some of the settings, we have a reasonably high number of features (at most 517). Useless features could have a severe impact on the performance of our learning algorithms. Thus, we also consider feature selection. We use forward selection to study which features contribute most to the predictive performance. We apply a simple decision tree classifier on the training data using the subset of features considered and measure its accuracy. Of course, the choice for the decision tree could make the selected features less suitable for the other algorithms but we want to keep things simple and therefore refrain from exploring this for every algorithm separately. Figure 7.11 shows the impact of adding the best features iteratively for the first 50 features. We see that after selecting around 5 features the performance

**Table 7.3** Attribute subsets used for classification case

| Dataset name | Basic features | PCA | Time-and frequency based features | Clusters | #features |
|---|---|---|---|---|---|
| Initial set (cleaned) | 21 | | | | 21 |
| Chapter 3 | 21 | 7 | | | 28 |
| Chapter 4 | 21 | 7 | 56 + 432 | | 516 |
| Chapter 5 | 21 | 7 | 56 + 432 | 1 | 517 |
| Selected features | | | 2 + 8 | | 10 |

**Fig. 7.11** Feature selection for the label prediction task (performance measured on the training set)

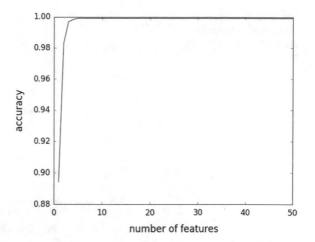

on the training set no longer substantially increases (by the way, look at the high accuracy scores we obtain in this initial run already!). We select 10 features as this number of features is selected based on a decision tree, other approaches might require slightly more features because they are able to exploit the additional information. The selected features include only time -and frequency based features. Time related features that are selected are the mean of the pressure measured by the phone and the standard deviation of the x-axis of the gyroscope. The frequency features included are the amplitude of the 0 Hz frequency (i.e. the intercept) for the y-axis of the accelerometer of the phone, the entropy of the y-axis of the magnetometer of the watch, the frequency with the highest amplitude for the z-axis of the magnetometer, the weighted frequency of the y-axis of the gyroscope of the watch, the amplitude of the 1 Hz frequency for the y-axis of the phone's gyroscope, the amplitude of the 1.9 Hz frequency of the x-axis of the accelerometer of the phone, and the amplitude of the 0.9 Hz frequency and 0.5 Hz frequency for the z-axis of the magnetometer of the watch and the y-axis of the accelerometer respectively.

### 7.10.1.3 Model Complexity and Tuning Parameters

We have identified the training set as well as the different subsets of features. We are almost ready to apply the various learning algorithms. Before that, however, we should be aware that we do not want to overfit our models towards the training data (remember our discussion in Chap. 6). Of course, one solution we have already considered, namely to select a number of features, but many alternatives exist. In this section we will focus on punishing overly complex models (regularization) and we should also set the parameters of the learning algorithms in an appropriate way.

Let us first consider regularization. As we have discussed we can add a regularizer to the objective functions, punishing more complex models. The regularizer comes with a regularization parameter. The higher the parameter value, the more complex

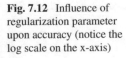

**Fig. 7.12** Influence of
regularization parameter
upon accuracy (notice the
log scale on the x-axis)

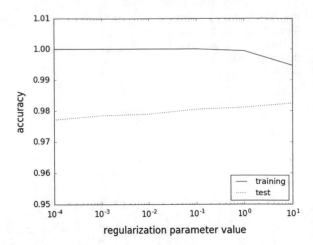

models are punished, and hence, the more simple models (with low weights) are favored. Figure 7.12 shows the impact of the regularization parameter upon the performance for both the training and the independent test set. Although the differences are not that big (this is very dependent on the variance of the data), we do observe a trend that the performance on the training set goes down (in fact it always behaves that way) when we increase the value of the parameter while the performance on the test set goes up. Hence, we end up with a better generalizable model thanks to an appropriate setting for the regression parameter.

Next, the setting of the parameter values of the learning algorithms can severely impact their performance. Some parameters represent the complexity of the resulting models as we have seen in the previous example, while others represent different aspects such as the shape of the objective function. We will focus on a decision tree again. One parameter allows us to set the minimum number of examples per leaf. If we set this number too low, overfitting is likely to occur since leafs represent only limited examples. Hence, we could say that this parameter says something about the complexity of the tree. If we consider Fig. 7.13 we see that performance on the training set increases when we decrease the minimum number of example per leaf. The same holds for the test set up to a certain extent. There is however a breaking point: the test set performance drops when a value smaller than 5 is selected, corresponding to overfitting.

So how do we find the best settings for these parameters? Well, we should refrain from using information from the test set (otherwise we are in fact optimizing on the independent test set and are snooping data). The approach we take is to try different parameter settings and evaluate their performance using a 5-fold cross validation approach on the training data. Using such an approach, we have a good indication on the generalizability of our models. We identify the most important parameters for each of the classification methods and determine a set of parameter values. We then

**Fig. 7.13** Setting for the minimum number of points per leaf for a decision tree learning algorithm versus the accuracy

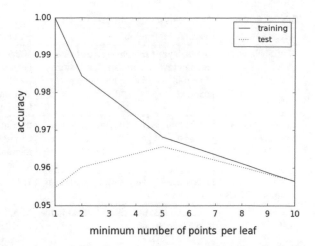

explore all different combinations (a grid search) and select the values that perform best in the cross validation. We use these to generate a model from the complete training set. Table 7.4 shows an overview of the algorithm and the variants as well as the parameters we vary. A lot of parameters can be varied but the ones we have selected provide sufficient opportunity to tailor the approach to the problem at hand. We will use all algorithms we have explained for classification (except the ones from the field of data stream mining).

### 7.10.1.4 Results

We now have all ingredients to run the algorithms and hopefully be impressed by their performance! We will evaluate the performance of the algorithms on both the training and test set. Various algorithms involve a stochastic process (random forests, neural networks, and the support vector machines depending on the kernel function) so for these we do not just do a single run of the algorithm but average their score across five runs of the algorithm. We could argue that it would be best to choose the single model with the highest value for the performance. However, we strive for an overall good performance, not for a single exceptional run.

The results are shown in Table 7.5. In the table we have also indicated the 95% confidence intervals. In order to calculate these intervals we compute the standard deviation of the accuracy ($a$) by $sd = \sqrt{a(1-a)/n}$. For the training set, $n = 1285$ while for the test set $n = 552$. The 95% confidence intervals are calculated by adding two times the standard deviation (upper bound) and subtracting it (lower bound). A graphical illustration of the performance on the test set is presented in Fig. 7.14. We can see that we obtain a very high accuracy on both the training and the test set. Apparently the problem is not very difficult to tackle with all of the sensory data we have. In addition, we only have data from several hours which might make

**Table 7.4** Algorithm variants and parameters varied

| Algorithm | Variant description | Parameters varied |
|---|---|---|
| Neural Network (NN) | Multi-layer perceptron with 1 or 2 hidden layers, a logistic activation function and one output node per class | Hidden layer composition: {single layer with 5, 10, 25, or 100 neurons, two layer with 100 and 5 neurons or 100 and 10} Maximum iterations: {1000, 2000} |
| Support Vector Machines (SVM) | SVM with kernel function with one SVM model per class | maximum iterations: {1000, 2000} C: {1, 10, 100} tolerance = {0.001, 0.0001} kernel function: {$rbf, polynomial$} |
| K-Nearest Neighbor (KNN) | KNN model using simple Euclidean distance | k: {1, 2, 5, 10} |
| Decision Tree (DT) | Decision tree algorithm following the CART approach | Minimum samples per leaf: {2, 10, 50, 100, 200} splitting criterion: {$gini, entropy$} |
| Naive Bayes (NB) | Basic Naive Bayes approach | - |
| Random Forest (RF) | Basic Random forest approach | Minimum samples per leaf: {2, 10, 50, 100, 200} number of trees: {10, 50, 100} splitting criterion: {$gini, entropy$} |

our problem easier because the circumstances under which the measurements have taken place can be assumed to be similar. Finally, we also have a substantial overlap between our windows in the time and frequency domain (remember that we selected a 90% overlap). This could have resulted in very similar instances in the training and the test set, making it easier to obtain good performance. If we would have had more data we could have reduced this overlap. In the literature related to activity recognition scores ranging between 0.65 and 0.98 have been observed [11, 78].

When we explore the results in a bit more detail, we observe that the addition of the temporal features is beneficial for some algorithms, especially for the random forest: we see a significant improvement as the confidence intervals do not overlap with those without the temporal data. The selection of features up front (i.e. taking the top 10 features) did not help: the performance is severely degraded across the algorithms except for the decision tree, this is probably caused by the fact that we use the decision tree algorithm to select the best features, biasing the selection. In terms of best performance, the random forest exhibits the highest score, although it is not significantly better compared to the other algorithms, except for the naive bayes approach. The k-nearest neighbor approach obtains the perfect score on the training set, caused by the fact that it simply memorizes all training data (hence, it is trivial to obtain the perfect score) but the generalizability is still good, this could be caused by the relative similarity of the data we train and test on.

**Table 7.5** Performance of algorithms on label classification task (NN = Neural Network, RF = Random Forest, SVM = Support Vector Machine, KNN = K-Nearest Neighbor, DT = Decision Tree, NB = Naive Bayes)

| Approach | NN | | RF | | SVM | | KNN | | DT | | NB | |
|---|---|---|---|---|---|---|---|---|---|---|---|---|
| Features | Training | Test | Training | Test | Training | Test | Training | Test | Training | Test | Training | Test |
| Initial set | 0.9894 (0.9837 - 0.9951) | 0.9699 (0.9554 - 0.9845) | 0.9991 (0.9974 - 1.0008) | 0.9674 (0.9523 - 0.9825) | 1.0000 (1.0000 - 1.0000) | 0.9710 (0.9567 - 0.9853) | 1.0000 (1.0000 - 1.0000) | 0.9728 (0.9590 - 0.9867) | 0.9914 (0.9863 - 0.9966) | 0.9275 (0.9055 - 0.9496) | 0.9300 (0.9157 - 0.9442) | 0.9130 (0.8891 - 0.9370) |
| Chapter 3 | 0.9897 (0.9841 - 0.9954) | 0.9670 (0.9518 - 0.9822) | 0.9988 (0.9968 - 1.0007) | 0.9659 (0.9505 - 0.9814) | 1.0000 (1.0000 - 1.0000) | 0.9710 (0.9567 - 0.9853) | 1.0000 (1.0000 - 1.0000) | 0.9728 (0.9590 - 0.9867) | 0.9899 (0.9843 - 0.9955) | 0.9257 (0.9034 - 0.9480) | 0.9214 (0.9064 - 0.9364) | 0.9094 (0.8850 - 0.9339) |
| Chapter 4 | 0.9981 (0.9957 - 1.0005) | 0.9743 (0.9608 - 0.9878) | 1.0000 (1.0000 - 1.0000) | 0.9938 (0.9872 - 1.0005) | 1.0000 (1.0000 - 1.0000) | 0.9855 (0.9753 - 0.9957) | 1.0000 (1.0000 - 1.0000) | 0.9783 (0.9658 - 0.9907) | 0.9946 (0.9904 - 0.9987) | 0.9656 (0.9501 - 0.9811) | 0.9479 (0.9355 - 0.9603) | 0.9167 (0.8931 - 0.9402) |
| Chapter 5 | 0.9989 (0.9971 - 1.0008) | 0.9725 (0.9585 - 0.9864) | 1.0000 (1.0000 - 1.0000) | 0.9928 (0.9855 - 1.0000) | 1.0000 (1.0000 - 1.0000) | 0.9855 (0.9753 - 0.9957) | 1.0000 (1.0000 - 1.0000) | 0.9783 (0.9658 - 0.9907) | 0.9953 (0.9915 - 0.9991) | 0.9601 (0.9435 - 0.9768) | 0.9471 (0.9346 - 0.9596) | 0.9167 (0.8931 - 0.9402) |
| Selected features | 0.8812 (0.8632 - 0.8993) | 0.8493 (0.8188 - 0.8797) | 0.9995 (0.9983 - 1.0007) | 0.9808 (0.9691 - 0.9925) | 0.9767 (0.9682 - 0.9851) | 0.9511 (0.9327 - 0.9694) | 0.8988 (0.8820 - 0.9157) | 0.8696 (0.8409 - 0.8982) | 0.9992 (0.9977 - 1.0008) | 0.9764 (0.9635 - 0.9894) | 0.8000 (0.7777 - 0.8223) | 0.8062 (0.7725 - 0.8398) |

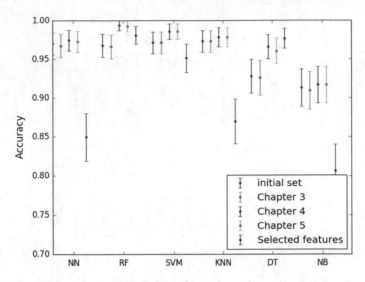

**Fig. 7.14** Visualization of accuracy include confidence intervals

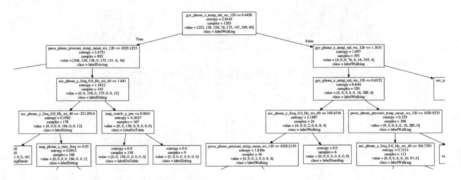

**Fig. 7.15** The resulting decision tree using the selected features

Let us dive into a bit more detail of the models that we have now obtained. The most insightful would be the decision tree. To simplify matters, we take the variant with the selected features. Figure 7.15 shows (part of) the decision tree. We see that the temporal feature which computes the standard deviation of the x-axis for the gyroscope of the phone is used in the root node. In addition, if we go down the tree we see that the frequencies of the classes in the training data (expressed as a vector following the identifier *value*) indeed move towards a single class. This happens very nicely, although the tree is rather specific and the leaves only have a limited number of examples. Still, the generalizability within this dataset is good. If we look into the details of the parameter settings, we see that a minimum of 2 cases per leaf is selected as well as the entropy splitting criterion.

Exploring one of the best classification algorithms, the random forest combined with the selected attributes, is a bit more tricky as we do not have a single tree but

**Table 7.6** Feature
importance Random Forest

| Feature | Importance |
|---|---|
| gyr_phone_x_temp_std_ws_120 | 0.2855 |
| press_phone_pressure_temp_mean_ws_120 | 0.2499 |
| acc_phone_y_freq_0.0_Hz_ws_40 | 0.2433 |
| mag_watch_y_pse | 0.0873 |
| mag_phone_z_max_freq | 0.0341 |
| gyr_watch_y_freq_weighted | 0.0284 |
| gyr_phone_y_freq_1.0_Hz_ws_40 | 0.0255 |
| acc_phone_x_freq_1.9_Hz_ws_40 | 0.0227 |
| acc_watch_y_freq_0.5_Hz_ws_40 | 0.0164 |
| mag_watch_z_freq_0.9_Hz_ws_40 | 0.0068 |

we have a lot of them (the best parameter value selected for the number of trees is 50 combined with a minimum of 2 samples per leaf and the entropy as a splitting criterion). We can however look at the importance of the various features, shown in Table 7.6.

Finally, we can explore where the random forest algorithm tends to make mistakes (i.e. which labels are confusing). The confusion matrix is shown in Table 7.16. We observe a lot of high numbers on the diagonal, indicating a high accuracy. We do see that the algorithm predicts a very limited number of cases incorrectly, washing hands is classified as eating two times, while walking is predicted to be standing for two times.

To summarize, we obtain excellent results for our dataset. However, it is hard to say whether we would perform equally well if we would have a richer dataset. Still, the steps that we have explained are precisely the same ones that you would perform for any other dataset, which is the main purpose of this case study.

## 7.10.2  Regression: Predicting the Heart Rate

The next task we are going to consider is to predict the heart rate of the user based on all other measurements, including the label. This is a regression task since we are aiming to predict a continuous value. We will address this as a temporal task: we will learn on the first part of the data (in terms of time) and predict for the remaining part of the data, following the setup we have described in Sect. 7.1. The approaches we use from this chapter will not exploit the ordering of the data, but we still have the temporal features we have identified in Chap. 4. We will go through each of the steps in somewhat less detail compared to our previous classification problem since the approach is very similar. Instead of the accuracy for our previous problem we will use the mean squared error as a performance metric (i.e. the lower the better).

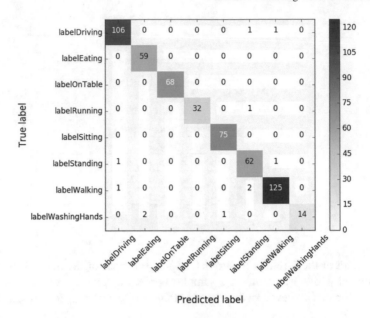

**Fig. 7.16**  Random forest confusion matrix for crowdsignals classification task

### 7.10.2.1  Preparing the Dataset for Learning

As said, we are going to address this problem as a temporal learning problem. We take an interval from the starting point of our measurements to about halfway (totaling to 1422 data points for training). We have selected this point because nearly all activities (that are likely to be highly influential on the heart rate) are reflected in this training period. As a test set we select the interval following our training interval. In the quantified self setting this would represent collecting some labeled data from the user at first, and seeing how well we are able to predict for that user. We only use part of the dataset for testing to avoid including totally new activities we did not train on (and hence, could not train our model properly for). Hence, we study how well we can create generalizable models for activities we have seen before, and not how well we generalize for totally new activities with very different measurement characteristics. This would be a more difficult problem while the problem is already pretty difficult as we will see. The test set contains 736 instances. Figure 7.17 illustrates the selected training and test set. In the regression case we do not remove any additional time points since the heart rate is known everywhere or has been interpolated at least.

### 7.10.2.2  Selecting Attributes

We again go through the attribute selection phase. We define a number of subsets similar to what we have done for the label prediction task. Table 7.7 shows the number

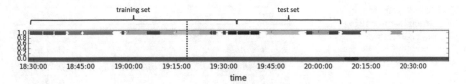

**Fig. 7.17**  Training and test set split for heart rate prediction task

**Table 7.7**  Attribute subsets used for regression case

| Dataset name | Basic features | PCA | Temporal features | Clusters | #features |
|---|---|---|---|---|---|
| Initial set (cleaned) | *21* | | | | 21 |
| Chapter 3 | *28* | *7* | | | 35 |
| Chapter 4 | *28* | *7* | *71 + 432* | | 538 |
| Chapter 5 | *28* | *7* | *71 + 432* | *1* | 539 |
| Selected features | *1* | *1* | *6 + 2* | | 10 |

of features included. We have a few more basic features (we clearly exclude heart rate, but include the binary attributes representing the individual activity labels). The same holds for the temporal features: the categorical temporal features for the labels have been included while the temporal features associated with the heart rate are no longer present.

We also made a selection of features, based on the Pearson coefficient. The correlation values for the selected features, which are the 10 features with the most extreme correlations, as shown in Table 7.8. When observing the correlations with the highest magnitudes, we see that the position of the phone on the table is quite important as are a few principal components. Note that the temporal feature for the phone lying on the table is slightly different from the regular label as it is true if it has been observed within the historical window. We again see that time and frequency based features play an important role.

### 7.10.2.3  Tuning Parameters

For tuning the parameters we take the same approach as we have considered before, namely do a grid search over parameter values that are potentially suitable. We again use cross validation since our learning algorithms do not care about the order anyway (this will change in the next chapter). An overview of the algorithms we use (again, all of those that we have explained and can be used for solving a regression problem have been included) and the parameter values that are part of the grid search are shown in Table 7.9. We use support vector regression without a kernel function due

**Table 7.8** Ten most important features using Pearson correlation

| Feature | Importance |
|---|---|
| *temp_pattern_labelOnTable* | 0.6158 |
| *labelOnTable* | 0.6158 |
| *temp_pattern_labelOnTable(b)labelOnTable* | 0.6158 |
| *pca_2_temp_mean_ws_120* | 0.5265 |
| *pca_1_temp_mean_ws_120* | 0.5221 |
| *acc_watch_y_temp_mean_ws_120* | 0.5022 |
| *pca_2* | 0.4980 |
| *acc_phone_z_temp_mean_ws_120* | 0.4970 |
| *gyr_watch_y_pse* | −0.5368 |
| *gyr_watch_x_pse* | −0.6419 |

**Table 7.9** Algorithm variants and parameters varied for the regression problem

| Algorithm | Variant description | Parameters varied |
|---|---|---|
| Neural Network (NN) | Mutli-Layer Perceptron with 1 or 2 hidden layers and an identity activation function and one output node per class | hidden layer composition: {single layer with 5, 10, 25, or 100 neurons, two layer with 100 and 5 neurons or 100 and 10} maximum iterations: {1000, 2000} |
| Support Vector Regression (SVR) | Linear SVR (no kernel function) | maximum iterations: {1000, 2000} C: {1, 10, 100} tolerance = {0.001, 0.0001} |
| K-Nearest Neighbor (KNN) | KNN model using simple euclidean distance | k: {1, 2, 5, 10} |
| Decision Tree (DT) | Decision tree algorithm following the CART approach with the mean squared error as a splitting criterion | minimum samples per leaf: {50, 100, 200} |
| Random Forest (RF) | Basic Random Forest approach with the mean squared error as a splitting criterion | minimum samples per leaf: {50, 100, 200} number of trees: {10, 50, 100} |

to the lengthy runtimes. We again average the performance of the random forest and the neural network over five runs.

### 7.10.2.4   Results

Table 7.10 shows the mean of the squared errors we have obtained for the time points in the test set, and the standard deviation over these squared errors as well. This gives an impression on the variability we see in our predictions over the different time points. Figure 7.18 shows the mean squared errors and their standard deviations for the test set. The results substantially differ from the classification task, for which

**Table 7.10** Performance of algorithms on heart rate regression task (NN = Neural Network, RF = Random Forest, SVR = Support Vector Regression, KNN = K-Nearest Neighbor, DT = Decision Tree)

| Approach | NN | | RF | | SVR | | KNN | | DT | |
|---|---|---|---|---|---|---|---|---|---|---|
| Features | Training | Test | Training | Test | Training | Test | Training | Test | Training | Test |
| Initial set | 725.3 (1009.3) | 1584.6 (1417.6) | 582.5 (857.1) | 1548.0 (1346.5) | 1155.8 (1216.8) | 1303.3 (1067.2) | 186.8 (574.1) | 2309.7 (2027.9) | 542.8 (855.8) | 1496.7 (1268.4) |
| Chapter 3 | 716.3 (971.6) | 1510.0 (1333.6) | 476.0 (800.9) | 1750.4 (1500.1) | 1600.2 (4785.8) | 1465.2 (1609.4) | 186.8 (574.1) | 2309.7 (2027.9) | 537.2 (850.8) | 1501.4 (1282.2) |
| Chapter 4 | 523.6 (846.2) | 1968.0 (2801.3) | 15.1 (41.6) | 1310.3 (1129.2) | 1516.1 (2783.7) | 1967.4 (3950.7) | 134.2 (473.6) | 2259.0 (1848.2) | 368.2 (643.8) | 1317.0 (826.8) |
| Chapter 5 | 557.7 (861.5) | 2014.7 (2114.9) | 7.0 (21.1) | 1384.8 (1292.9) | 1211.8 (2051.6) | 1997.2 (3243.5) | 134.2 (473.6) | 2259.0 (1848.2) | 368.2 (643.8) | 1317.0 (826.8) |
| Selected features | 768.9 (1369.2) | 2825.3 (2855.4) | 620.7 (1109.3) | 2614.8 (2613.4) | 808.5 (1909.1) | 2598.2 (2944.2) | 286.1 (719.9) | 2571.1 (2896.0) | 538.6 (1039.8) | 2740.0 (2878.4) |

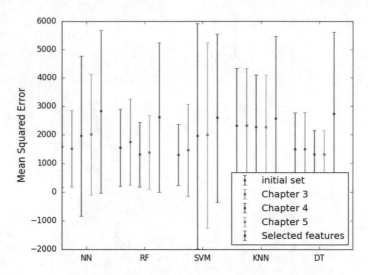

**Fig. 7.18** Performances (mean squared error) on test set, including standard deviation

we performed really well. We observe that some algorithms are able to predict the heart rates (random forest and the k-nearest neighbor) with a very low error on the test set (though the standard deviations are generally very high, caused by several very wrong predictions). The generalizability is unfortunately very limited when we look at the performance obtained on the test set. Hence, we overfit towards the training data. Given the consistent lack of performance this likely caused by an insufficiently representative training set. Additionally, external influences of the heart rate might not be represented in our features. If we look at the figures of our dataset we have shown earlier we can in fact see this. For instance, for the walking activity in the training set we see a reasonably low heart rate while we see an elevated heart rate in the test set. The other sensory values that are used as a predictor are, however, quite similar over both periods. Hence, our model cannot predict this well. When we study what features are most predictive we observe no consistent patterns over the different algorithms.

Let us dive into one of the approaches in a bit more detail. We select the random forest with the features from Chap. 5 (good performance on training and test set). The best parameters found are 10 trees, and a minimum of 10 samples per leaf in the tree. The key predictors used in the random forest are shown in Table 7.11. We see very diverse predictors, although the temporal predictors do play quite a dominant role. Figure 7.19 shows the performance of the random forest on both the training and test set. We see that it is able to reproduce the training set quite accurately while the predictions on the test set are indeed pretty bad, although some of the sudden changes are predicted correctly. Ideally, we would have more data over a longer period, although we would still expect the prediction to be quite difficult due to the huge variability in the heart rate in pretty similar situations.

**Table 7.11**  Feature importance Random Forest

| Feature | Importance |
|---|---|
| *acc_watch_y_temp_mean_ws_120* | 0.4631 |
| *press_phone_pressure_temp_mean_ws_120* | 0.1398 |
| *mag_watch_x_temp_mean_ws_120* | 0.0465 |
| *temp_pattern_labelEating(b)labelEating* | 0.0404 |
| *mag_watch_y_temp_mean_ws_120* | 0.0385 |
| *temp_pattern_labelEating* | 0.0383 |
| *labelEating* | 0.0382 |
| *pca_4_temp_mean_ws_120* | 0.0208 |
| *pca_7* | 0.0131 |
| *pca_7_temp_mean_ws_120* | 0.0112 |

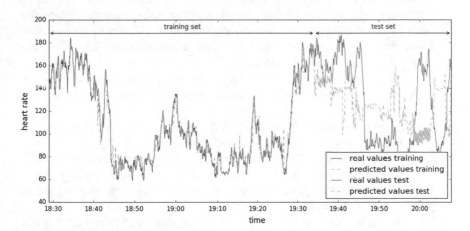

**Fig. 7.19**  Performance training and test set best performing model for regression task

## 7.11  Exercises

### 7.11.1  Pen and Paper

1. Imagine that we have a dataset with instances covering measurements that have been obtained using different mobile phones. What would your learning setup be? Will you just consider the dataset as a whole, or would you split it up? Take the learning setups we have considered into account.
2. Give an overview of the complexity of the following learning algorithms we have presented in this chapter: neural networks, support vector machines (with and without kernel), k-nearest neighbor, decision trees, and naive bayes. Do this

for the training process for a classification problem and express the complexity in terms of the number of instances $N$ and the number of features $p$.

3. Setting the parameter values of the learning algorithms properly is an important step to obtain good results (we have seen that in our case study), but trying out a lot of values might be very time consuming. Sometimes some rough guidelines are however available. Try to find a guideline on the number of neurons to use for a simple multi-layer perceptron network.

4. While back propagation is a very popular algorithm to train a neural network, it does come with some severe problems. List two disadvantages of the algorithm and provide a solution to each of these problems from the literature.

5. It is known that normalization of the input and output of the neural network is generally a good idea (although we did not fully do it in this case), explain why.

6. In our explanation of support vector machines we mentioned the existence of kernel functions. Also in deep neural networks and other approaches these are commonly seen. List at least three kernel functions and explain their properties.

7. Support vector machines are currently the algorithm of choice in the domain of text mining, why would this be the case? (hint: think of the number of attributes we are likely to face with text mining).

8. If we have lots of features (i.e. $p$ is large) nearest neighbor approaches tend to perform relatively poor compared to model-based approaches. Explain why this is the case.

9. Decision trees are known to be prone to overfitting. List and explain three approaches that could help us overcome this problem.

10. In our explanation of decision trees we have focused on the information gain for classification problems. There are however ample alternative splitting criteria. Find at least two other splitting criteria for a decision tree used for classification problems. Explain in detail how they work.

11. Naive Bayes thanks its name to the naive assumption that features are independent. Provide a concrete example in the context of the quantified self where this assumption is clearly violated.

12. Assuming that we would have the perfect training set for a specific problem which provides us with rich enough information to generate a highly generalized model on, would we need the concept of bagging?

13. Feature selection can be very useful. Explain how feature selection relates to overfitting.

## 7.11.2   Coding

1. In our setup for the activity recognition within our crowdsignals dataset we have ignored the cases that either contained two labels or had an unknown label: we simply removed them from our dataset. Experiment with the dataset in case we would not throw these out, what is the impact on the performance?

2. For now we have studied a simple learning setting with data of just one person. Consider the dataset you have previously identified covering multiple individuals. We assume that we are going to learn to predict unseen data of known users. We could do this in two ways: (1) generate separate models for each individual on the training data of that individual and predict the unseen data, or (2) generate a single model across the training data of all individuals and use that to predict the unseen data of each individual. Implement the two approaches for your own dataset, summarize the results, and draw conclusions on which approach is most suitable for your specific dataset.

3. Similarly to what we have done for the crowdsignals dataset, apply the learning algorithms to the dataset you have collected yourself and predict the activity. Compare the results in terms of accuracy. Do you find the same learning algorithms on top? And how do the results compare to what we have found for the crowdsignals dataset?

4. Make your own implementation of a Hoeffding tree which we have explained in this chapter. Compare the computational time required to run the algorithm and the resulting tree to a "standard" decision tree algorithm.

5. Compare backward to forward selection using one of your datasets. Are the same attributes selected for the two approaches? If there are major differences, can you explain why these differences occur?

# Chapter 8
# Predictive Modeling with Notion of Time

In our previous chapter we have looked at a variety of classification and regression algorithms. The approaches did however not consider the notion of time explicitly, which is a shame, because there might be valuable information. Thinking of Bruce, his past mood ratings over the last days might together be very predictive for his mood tomorrow, or for Arnold his progress during the past days could be indicative for his performance tomorrow. Although we were able to derive temporal features from our original sensory data, it might be much more natural to take these temporal aspects into account in the learning algorithm explicitly, right? Well, you are in luck because we are going to discuss these temporal learning algorithms in more depth in this chapter.

Consider Fig. 8.1. We see the mood of Bruce plotted over time together with the value for the activity level. We see that a rise in activity level results in delayed increase in the mood level. Furthermore, we see that certain days of the week generally have higher mood and activity levels than others. Let us look at the approaches we can use to predict these types of patterns, but not before we again discuss the learning setup.

## 8.1 Learning Setup

For this case, we can use a learning setup which is very much like the one we described in the previous chapter (in Table 7.1). Of course, in this case we only consider the temporal column (i.e. we assume an $\mathbf{X}^{\mathcal{T}}$). Most of the models we will treat assume a single temporal sequence of values to learn from (i.e. they focus on the individual level). It is however possible to feed multiple temporal sequences and minimize the error across all those temporal sequences.

© Springer International Publishing AG 2018

M. Hoogendoorn and B. Funk, *Machine Learning for the Quantified Self*,
Cognitive Systems Monographs 35, https://doi.org/10.1007/978-3-319-66308-1_8

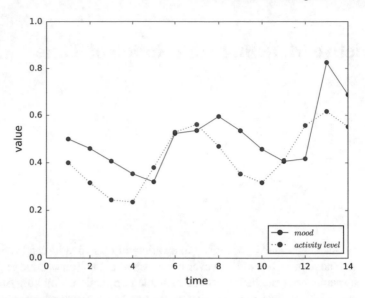

**Fig. 8.1** Example dataset Bruce visualized

## 8.2  Time Series Analysis

Classical time series analysis [31, 34, 107] provides tools that can help us to find regularities and trends in temporal data. For example, we might observe Bruce's mood to vary periodically over the course of a week. On Monday and Tuesday Bruce is always somewhat depressed as many people are, while his mood improves over the following weekdays. At the same time, Bruce is continuously improving his mental health through smartphone based E-Mental Health interventions. So, in addition to the weekly pattern Bruce's mood show a long term increase.

Temporal patterns can be analyzed on different time scales. While the previous example has a time scale of days or weeks, periodicity can occur on much shorter time scales. Think for example of Arnold doing his favorite track—the accelerometer data will exhibit a strong periodicity with a frequency around 1 or 2 Hz. And there are even shorter time scales, for example if we want to analyze sleeping patterns using the microphone of a smartphone periodicities occur at frequencies between 1 and 5000 Hz [56]. In Sect. 4.4.2 we explored the Fast Fourier Transformation which analyzed the spectral components of a signal in the frequency domain. In contrast, classical times series analysis studies a signal in its time domain.

In general, the goals of time series analysis fall into one of the following categories: (i) understanding periodicity and trends, (ii) forecasting, and (iii) control (i.e. change the course of a temporal pattern). Even though time series analysis is a well elaborated and rich field (and to summarize the concepts in one section is therefore challenging), we believe that it is valuable for the analysis of quantified self data to have a basic understanding of the concepts and tools.

## 8.2.1  Basic Concepts

Time series and their variations can be decomposed in three components: First, *periodic variations* might be due to a daily, weekly, monthly, or annual seasonality as discussed in the mood example of Bruce. Periodic variations can also result from other underlying mechanisms, e.g. the physics of running as in the example of Arnold. Second, a long-term *trend* in the data describes how its mean evolves over time. As you can imagine, depending on the specific context, it has to be defined what long-term means and what form a trend might have—it does not always have to be linear. Third, after removing the periodic component and the trend we are left with *irregular variations*. Residuals can be due to random noise or some other unknown and irregular causes.

This leads us to an important characteristic of a time series: stationarity. It is beyond the scope to formally introduce stationarity as a probabilistic concept.[1] We refer the interested readers to text books such as the one by Chatfield [34]. Here, we choose a more intuitive and data-driven approach to understand stationarity. We call a time series stationary if (i) trends and periodic variations are removed and (ii) if the variance of the remaining residuals is constant over time. This means that both the expected mean of a time series as well as its variance are constant. As we will see stationarity is a major prerequisite (or intermediate step) for many methods in time series analysis.

Let us look at a univariate time series $x_t^j$. For the sake of a concise notation we will drop $j$ in the following consideration - so keep in mind, that in this section $x_t$ does not refer to a vector with values for all attributes, but is the time series for one of the attributes.

In a more general definition of stationarity it is not only the variance that has to be constant over time but also the lagged autocorrelation which can be defined as

$$
r_\lambda = \frac{\sum_{t=1}^{N-\lambda} (x_t - \bar{x})(x_{t+\lambda} - \bar{x})}{\sum_{t=1}^{N}(x_t - \bar{x})^2} \tag{8.1}
$$

where $\bar{x}$ is the sample mean of the time series $x_t$. Autocorrelation with lag $\lambda$ measures (as the name suggests) the correlation between a time series and a shifted variant (by $\lambda$ time steps) of itself. It is an important measure that can provide clues to an underlying model that describes the data well and characterizes the predictability of a time series. Figure 8.2 shows three example time series and the corresponding empirical autocorrelation plots, also called correlograms. Figure 8.2a represents random noise, where each data point is independent and identically distributed from the same normal

---

[1] It should be noted that it is not the time series itself that is stationary, but an underlying probabilistic model that is assumed to generate the time series. Nevertheless, in practice the term stationarity is often tagged to a time series. See Sect. 8.2.3.

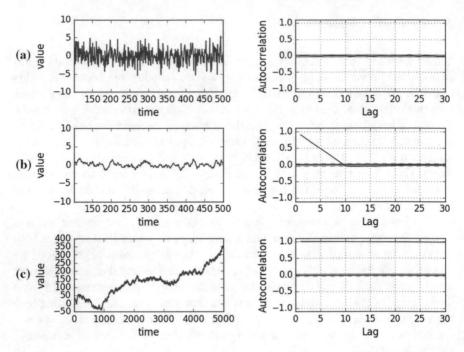

**Fig. 8.2** Lagged Autocorrelation for various time series: **a** random, **b** moderate autocorrelation, **c** non-stationary

distribution. As you would expect there is no autocorrelation for $\lambda > 0$. In Fig. 8.2b we introduce some memory over a couple of time steps which leads to autocorrelation. How we can calculate values using such a memory will be discussed in the next sections. Finally, Fig. 8.2c displays a time series that is generated by calculating the cumulative sum of the time series under (a). Because time series (c) exhibits a linear trend, you see immediately that it is non-stationary. In the next section we learn how we can model trends and periodicity of time series.

## 8.2.2  Filtering and Smoothing

To analyze trends in time series, we start with a simple approach that is close to what we have already seen in Chap. 2. Remember, what we did in order to reduce the noise of the raw data: We defined a regular spaced time grid with a time window $\Delta t$ and aggregated the unevenly spaced (raw data) measurements accordingly ($\mathbf{X}^{\mathcal{T}}$). As a result we were able to not only reduce the variability of data and to handle missing values but also to bring the data into a rectangular shape as it is useful for a variety of machine learning algorithms discussed in Chap. 7.

We can generate a new time series $z_t^j$ that is a linear transformation of an existing time series (for instance the point $x_t^j$), by defining

$$z_t = \sum_{r=-q}^{q} a_r x_{t+r} \tag{8.2}$$

where $a_r$ is a weight vector and $q$ is the number of measurements in each direction from $t$ that are taken into account when generating the new time series $z_t$. Such a transformation is called a linear filter. Linear filters are both useful and powerful. Let us look at a simple case where $a_r = (2q + 1)^{-1}$ for $r \in \{-q, \ldots, q\}$ and $a_r = 0$ for all other $r$. As a result $z_t$ is the average of the $2q + 1$ adjacent measurements of $x_t$, the so called *moving average*. If we believe that the measurements close to $t$ are more important for the newly generated time series we can simply increase the weights around $t$. For example we could give $a_r$ a triangular shape:

$$a_r = \begin{cases} \frac{q-|r|}{q^2} & -q \leq r \leq q \\ 0 & otherwise \end{cases} \tag{8.3}$$

Another common approach is exponential smoothing

$$a_r = \frac{\alpha(1-\alpha)^{|r|}}{2-\alpha} \tag{8.4}$$

where $\alpha$ is a constant with $0 < \alpha < 1$. The smaller we choose $\alpha$ the more importance is given to past and future measurements (Fig. 8.3). Often exponential smoothing only takes past measurements into account (enables online processing of data streams).

Do you remember what we wanted to do? Right, find and remove a trend. An effective way to do so, is to use a technique called differencing. Differencing is a special and at the same time simple filter with $a_0 = 1$ and $a_1 = -1$ and all other $a_r = 0$ which results in $z_t = x_t - x_{t-1} = \nabla x_t$. Here, $\nabla$ is defined as the operator that calculates the difference between two adjacent measurements. If we apply the $\nabla$-operator multiple times, e.g. $d$ times, we say that we apply $d$-th order differencing, e.g.

$$\nabla^2 x_t = \nabla x_t - \nabla x_{t-1} = x_t - 2x_{t-1} + x_{t-2} \tag{8.5}$$

To see why differencing is able to remove trends, we have to understand that the contribution of a long-term trend to two adjacent measurements is quite similar. Thus, subtracting the two measurements removes the trend. A drawback of differencing is that the variance of the detrended time series increases. This is true because $x_{t-1}$ is not a very good approximation of the trend component. However, we can interpret $z_t$ as a proxy for the trend. Thus, after taking the difference $x_t - z_t$ only the local, e.g. periodic, variation remains. This corresponds to a new linear filter with weights $b_r = -a_r$ for all $r$ except $r = 0$ for which $b_r = 1 - a_r$. We apply this by using our exponential smoothing filter (Fig. 8.4).

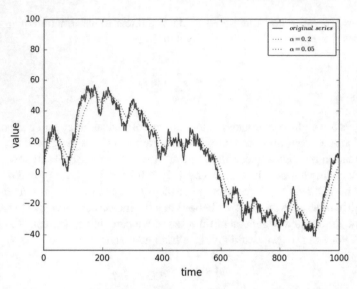

**Fig. 8.3** Exponential Smoothing of the time series from Fig. 8.2. *Black* dotted (*red* dotted) line corresponds to $\alpha = .2$ ($\alpha = .05$)

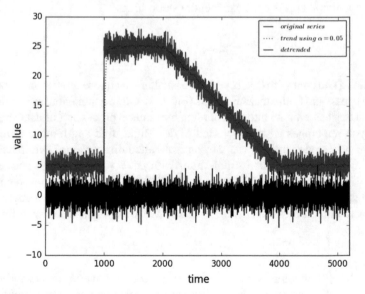

**Fig. 8.4** *Black* solid line = example time series, *red* dashed = trends through exponential smoothing, *blue* dotted line = detrended time series

"Art" is a term we use in the subtitle of this book—here we have another example: it is a kind of an art to say what the $q$ and $a_r$ should be to find trends, it depends on our understanding of the problem at hand. For example, if you want to use a smartphone to detect stepping patterns of person that has her phone in the pocket, there might be periodic pattern with a frequency of about 1 Hz. At the same time the orientation and position of the smartphone in the pocket of that person might vary over longer time intervals. To extract the stepping pattern, we would like to remove changes that are due to the changing orientation of the smartphone, so we would use a time interval for smoothing that is significantly larger than 1 s. We will return to this in an example later.

### 8.2.3 Autoregressive Integrated Moving Average Model—ARIMA

In this section we will take a more conceptual approach to modeling time series. We do so because we ultimately want to estimate a model that describes the empirical data well and that can be used to predict future data points.

We can think of time series as being generated by a stochastic process, that is a mapping of time point $t$ to a probability distribution $P_t$. It is this probability distribution that is assumed to generate the actual measurement $x_t$. Based on this distribution we can define the *mean function*

$$\mu(t) = E[P_t] \tag{8.6}$$

and the *autocovariance function*

$$\gamma(t_1, t_2) = E[(P_{t_1} - \mu(t_1))(P_{t_2} - \mu(t_2))] \tag{8.7}$$

Following up on our discussion along the empirical Eq. 8.1 we say a stochastic process $P_t$ is stationary, if the mean function $\mu(t) = \mu$ is constant over time and the autocovariance function only depends on the time difference $\lambda = t_2 - t_1$

$$\gamma(t_1, t_2) = \gamma(\lambda) \tag{8.8}$$

Let us start with a discrete random process for which $W_t$ (sometimes called white noise, hence the symbol $W$) is the normal distribution with variance $\sigma_W^2$ and all the observed data points are independently drawn from this distribution (independent and identically distributed, *i.i.d.*). Since the observations are statistically independent, we know that only the expected variance is non-zero while all other autocovariance terms are zero.

$$\gamma(\lambda) = \begin{cases} \sigma_W^2 & \lambda = 0 \\ 0 & otherwise \end{cases} \tag{8.9}$$

This situation is shown in Fig. 8.2a. Based on the purely random process we can define what is called a random walk process $P_t$ that follows

$$P_t = P_{t-1} + W_t \tag{8.10}$$

Generating data based on this model leads to a drift of the observations drawn from $P_t$. Neither the mean function nor the variance is constant over time as you can guess from Fig. 8.2c. This is why the stochastic process of a random walk cannot be stationary.

Another important stochastic process is the moving average (MA) process. Basically it applies a sliding window of fixed size $q$ and calculates the weighted mean of adjacent measurements for every $t$. Consider a purely random process $W_t$ as discussed above. Then

$$P_t = \theta_0 W_t + \ldots \theta_q W_{t-q} \tag{8.11}$$

is an MA process of order $q$ and with constant coefficients $\theta_i$. We can simplify Eq. 8.11 by using the backshift operator $B$ that acts on $W_t$ as follows: $W_{t-1} = BW_t$ which in general leads to $W_{t-n} = B^n W_t$. With $\theta(B) = \theta_0 + \theta_1 B + \ldots \theta_q B^q$ Eq. 8.11 simply becomes $P_t = \theta(B)W_t$

An MA process is shown in Fig. 8.2b. As you can see, there is clearly autocorrelation between adjacent measurements, e.g. $P_t$ and $P_{t-1}$, which is not surprising given that in this case $q$ terms of the sum in Eq. 8.11 are identical (not accounting for different $\theta_i$). It can be shown that consistent with Fig. 8.2b the autocovariance function is

$$\gamma(\lambda) = \begin{cases} 0 & \lambda > q \\ \sigma_W^2 \sum_{i=0}^{q-\lambda} \theta_i \theta_{i+\lambda} & 0 \le \lambda \le q \end{cases} \tag{8.12}$$

Knowing the autocovariance function $\gamma(\lambda)$ already tells us a lot about the MA process. E.g. the cut-off happens at lag $q$ which is the order of the MA process. However, as you can explore in the exercises, different MA processes can have the same autocovariance function. So, it is not enough to know the autocovariance function in order to unambiguously find the underlying time series model.

Next we want to look at *autoregressive processes (AR)*. If $W_t$ is a purely random process then an AR process follows

$$P_t = \phi_1 P_{t-1} + \ldots \phi_p P_{t-p} + W_t \tag{8.13}$$

where $\phi_i$ are constants (with $\phi_p \neq 0$). Using the backshift operator we can elegantly rewrite Eq. 8.13 as

$$\phi(B)P_t \equiv (1 - \phi_1 B - \cdots - \phi_p B^p)P_t = W_t \qquad (8.14)$$

Even though it takes some careful mathematical considerations, you can intuitively imagine that Eq. 8.14 can be rewritten as

$$P_t = \phi^{-1}(B)W_t = (1 + \theta_1 B + \theta_2 B^2 + \ldots)W_t \qquad (8.15)$$

which can be proven using the Taylor expansion of $(1 - \phi_1 B - \cdots - \phi_p B^p)^{-1}$. This is an interesting representation of an AR process because it shows that an AR process of finite order $p$ can be represented as an infinite order MA process (compare with Eq. 8.11). The opposite is also true for a finite order MA process—it can be represented as an infinite order AR process. This is referred to as the duality of AR and MA processes [95].

Why is that interesting? Since AR processes can be represented as MA processes and vice versa, it would be sufficient to only use one of them to fully describe a process that originally was generated by a mixture of both an AR and a MA process. However, we are interested in finding a process model that is as simple as possible in terms of the number of unknown parameters (principle of parsimony). For this purpose we mix an AR process of order $p$ and a MA process of order $q$ into an ARMA(p,q)

$$P_t = \phi_1 P_{t-1} + \ldots \phi_p P_{t-p} + W_t + \theta_1 W_{t-1} + \ldots \theta_q W_{t-q} \qquad (8.16)$$

When we turn to real time series, e.g. the mood of Bruce, we often observe that they are not stationary and thus, cannot be modeled by an ARMA process. One reason for that can be a drift in the mean. In the previous section we applied differencing to remove drifts of the mean. That is what the *autoregressive integrated moving average (ARIMA)* model does by replacing $P_t$ in Eq. 8.16 by $V_t = \nabla^d P_t = (1 - B)^d P_t$. The term "integrated" comes from the fact that $V_t$ has to be summed up to yield the original, non-stationary, time series $P_t$. The order of differencing $d$ enables us to not only remove linear trends but also trends that can be approximated by higher order polynomials. We say ARIMA-models are of order $(p, d, q)$.

ARIMA models can be extended to also account for seasonal variations, e.g. the weekday dependency of Bruce's mood. The basic idea behind it is not to apply the differencing to neighboring measurements but to those measurements which are separated by the seasonal period, in this case 7 days. In addition, we can include latent variables (our attributes) to predict the values of our time series in combination with the techniques we have discussed in the section. This is called the ARIMAX model. It is beyond the scope of this book to discuss the details of seasonal ARIMA and ARIMAX models (see the seminal textbook from Box and Jenkins [26]).

With ARIMA processes we have a powerful instrument to describe models that generate univariate time series data. In the next section we will provide you with basic ideas how to estimate ARIMA models.

## 8.2.4  Estimating and Forecasting Time Series Models

Let us assume we know that a time series $x_t$ was generated by an AR process of order $p$ with mean equal to 0. Then we can use Eq. 8.13 to find least square estimates for $\phi_1, \ldots, \phi_p$ by minimizing

$$S = \sum_{t=p+1}^{N} (x_t - \phi_1 x_{t-1} - \cdots - \phi_p x_{t-p}) \tag{8.17}$$

We refrain from going through the details of the solutions for specific values of $p$ as they can be found in dedicated textbooks (see for example p. 121 and further in [107]). But one important question is still open: How do we find the right value for order $p$? To answer this question, the so-called *Partial Autocorrelation Function (PACF)* is used. Basically, the PACF of order $p$ measures the correlation between two measurements $x_t$ and $x_{t+p}$ that is not already captured by the AR(p-1) process [34]. That said, the empirical PACF plot of a time series that was generated by an AR(p) process cuts off at lag $p$ and therefore presents a clue to what the order of the AR process is.

When it comes to fitting a pure MA process, finding the right order $q$ of the process is reasonably easy: As we have argued before, the *Autocorrelation Function (ACF)* cuts off at lag $q$. At the same time, estimating the parameters of the generating MA process is more difficult and found by employing numerical techniques. The same is true for ARMA (and ARIMA) processes. To estimate a non-seasonal ARIMA process a grid search over the 3 parameters $(p, d, q)$ can be performed and evaluated along typical criteria for model selection such as *Akaike's Information Criterion (AIC)* or *Bayesian Information Criterion*, accounting for both, the goodness of fit and the number of independent parameters. We will explore this in the case study.

We engage in time series analysis to not only understand the different components that are likely to have generated the data we observe but also to predict what happens in the near future. Once we have estimated the parameters of an ARMA process we can use Eq. 8.16 to predict future measurements. Due to the white noise $W_t$ the prediction quickly becomes uncertain and approaches the mean of the time series under consideration. In addition to that we have to account for seasonal trends in the data. As we will see in the case study, this can be tricky.

A concluding remark before we jump to a concrete example: time series analysis is an active field of research and extensively used in practice. So, there is a wealth of other topics related to time series analysis that are covered in other chapters and sections of this book such as Kalman Filtering (3.3), Fourier Transformation

(4.2.1), and to some extend state-space-models (8.4). For other topics that are equally important for time series analysis such as multivariate time series (VARMA model), long-memory models, or Bayesian analysis of time series we refer to standard text books (e.g. [31, 34, 43, 107]).

## 8.2.5 Example Application

In order to understand the application of the times series approach we have discussed, let us look at an example. We focus on finding the orientation of a smartphone based on the accelerometer data and predicting future accelerometer data.

### 8.2.5.1 Finding the Smartphone Orientation

We have explored different techniques of filtering which can be useful for differencing and smoothing. Figure 8.5 demonstrates the effect of a simple filter ($a_0 = 1$ and $a_1 = -1$). The left/right panel show the original/filtered data (granularity of 50 ms). The de-meaning of the data comes with a price: It increases the variance of the data.

Let us use filtering to get a rough estimate of the orientation of the smartphone. This can be done because gravitation introduces a natural coordinate system that is detected by the smartphone's accelerometer sensors. To do so, we apply a filter ($a_{-100} = \cdots = a_{100} = 1/201$) to the accelerometer data (along the three axis $x$, $y$, $z$) that is a sliding window over 201 adjacent measurements. As a result, this sliding window calculates the mean at each point in time. While the grey line in Fig. 8.6 corresponds to the original data (50ms granularity) and exhibits significant variability, the blue line represents the filter result. The red line indicates the norm of the 3D acceleration vector. If we would change the orientation of a smartphone slowly

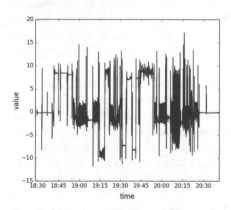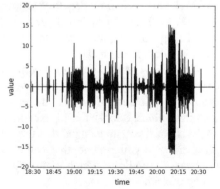

**Fig. 8.5** Differencing accelerometer data with a simple filter ($a_0 = 1$ and $a_1 = -1$)

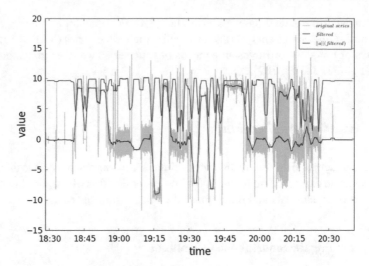

**Fig. 8.6** Unfiltered and filtered accelerometer data

we would expect the red line to equal the gravitational acceleration of the earth ($g \approx 9.81 \, \text{m/s}^2$). This is roughly the case. Deviations are due to several reasons: (i) rotational movements can increase the norm of the acceleration vector, (ii) free fall decreases it (typically not a good state for a smartphone and/or its owner, especially if you think of the time window size), and (iii) measurement errors (see for example at time 21:10, this decrease is due to a cut off at 20 m/s$^2$ in the raw data).

### 8.2.5.2  Predicting Accelerometer Data

Let us turn the application of a seasonal ARIMA model. We start by analyzing a fragment of 4000 accelerometer data measurements ($\approx$20 s) which we evenly space at a 10 ms level to accommodate for a proper time series application. The corresponding ACF (Fig. 8.7) exhibits a slowly decaying periodic autocorrelation. Together with a visual inspection of the raw data (Fig. 8.8) we see a strong seasonal component without a drift in the mean.

We want to ignore the seasonal component first and fit the best possible ARMA(p, q) process by applying a grid search over $(p, q)$. The objective criterion is to minimize the AIC. This results in $p = 3$ and $q = 2$. Figure 8.8 shows the initial time series and the fitted values. It looks like an exceptional result which you often see in time series analysis, until you understand that the fitted values are the so-called one-step-ahead prediction. So you assume, that you have all measurements up to a time $t - 1$ and you want to predict $x_t$ which intuitively should be easy if the time series is smooth enough which it is.

For some control tasks that might be a helpful result. However, typically we are more interested in long-term prediction based on the estimated model. Figure 8.9

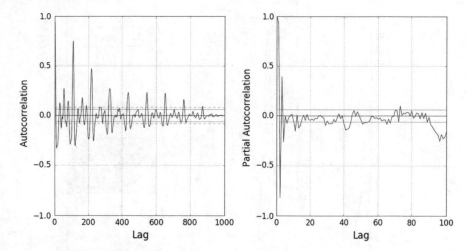

**Fig. 8.7** Autocorrelation Function (ACF) for a set of 2000 raw data measurements ($\approx 10$ s) and Partial Autocorrelation Function (PACF) for the same set of raw data

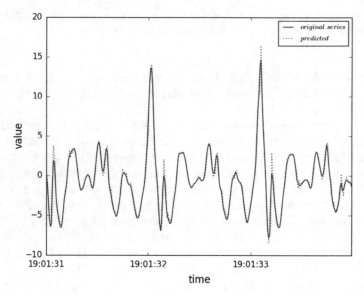

**Fig. 8.8** The *blue line* represents 500 measurements of the original data, the *red line* the one-step-ahead-prediction

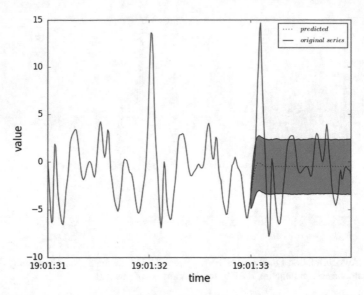

**Fig. 8.9** The *blue line* indicates the original time series that was used to estimate the ARMA model. The *red line* is the "long-term" prediction, the shaded area its uncertainty

shows that we do not do very well on this task. Already, after a few time steps the prediction uncertainty increases dramatically and we are way off, we basically predict the mean.

So, what went wrong? As we said before, the original time series exhibits a strong seasonality which has to be accounted for when fitting the model, here a *seasonal ARIMA (SARIMA)* model. We can decompose our time series in a seasonal component and a residual part (difference between the measured values and the seasonal component). The ARIMA model can then focus on the residue. If we look at Fig. 8.8 we have seen before, we observe a recurrent pattern about every second (1.07 s to be precise). Figure 8.10 shows the decomposition. We can see that the seasonality component accounts for a large part of our data.

We again take $p = 3$ and $q = 2$. As we can see in Fig. 8.11 we have a decent fit. If we want to predict measurements that are even further in the future, again the prediction levels are off and the interval of uncertainty increases. The reason for that is, that the long-term lagged autocorrelation decreases with the lag, so that predictions become more and more uncertain. Overall, you can see that the instruments we have described are very helpful when it comes to predicting time series with some regularity.

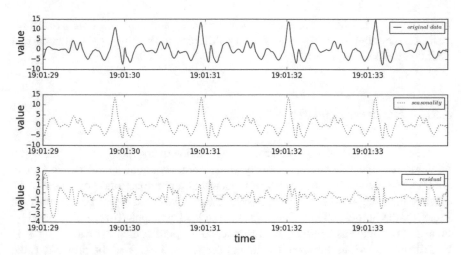

**Fig. 8.10** The data decomposed in a seasonality and a residual part

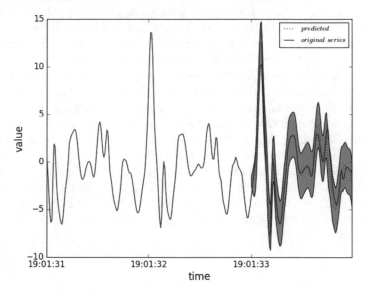

**Fig. 8.11** The *blue line* indicates the original time series that was used to estimate the SARIMA model. The *red line* is the "long-term" prediction, the shaded area its uncertainty

## 8.3 Neural Networks

We have already seen several variants of neural networks. Here we will focus on neural networks that can explicitly deal with temporal data—the so-called *recurrent neural networks*. We will discuss two types: the most basic variant that underlies most popular recurrent neural network approaches and *echo state networks*. We describe the latter because they are very different from other types of recurrent neural networks.

### 8.3.1  Recurrent Neural Networks

Figure 8.12 shows an example of a recurrent neural network. Compared to the neural network examples we have previously considered, recurrent neural networks can contain cycles that feed the values of a neuron at a previous time point (assuming discrete time steps) back into the network. Hence, we create a form of memory. In Fig. 8.12, the output of the network $(Y_1)$ is fed back as input. Time is expressed by means of the brackets behind each attribute, so we predict the next value $Y_1(t + 1)$ based on the previous values of the input $(X_1(t), \ldots, X_p(t))$ and the previous value of the output $(Y_1(t))$. Precisely what values are fed back into the network depends on design choices. In the network at hand it only concerns the output, but also values of hidden neurons could be fed back to the input or to another hidden neuron. Let us consider the example shown in Fig. 8.1 again. We might want to predict the mood of Bruce based on the activity level. We can build a recurrent neural network for predicting mood at the next point in time (i.e. $t + 1$). As input for the model, we use the previous mood prediction by the network $(Y_1(t))$ and the other measurements at the previous time point $(X_1(t), \ldots, X_p(t)$, in this case only activity level).

A crucial question is how we train these networks. We have seen the *backpropagation* algorithm before, but that does not take the notion of time into account. With this new setting our predictions not only depend on the inputs of the network but also on the values of neurons at the previous time step. As a consequence, we cannot update the weights based on the observed error by solely looking at the state of a network at a single time point. So how do we solve this problem? Well, it works less complex as you might expect: we "unfold the network through time", this means

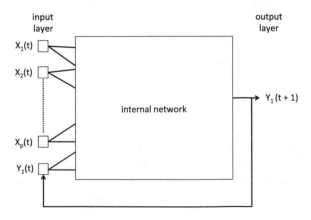

**Fig. 8.12** Simplified recurrent neural network

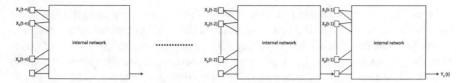

**Fig. 8.13**  Unfolded recurrent neural network

that we create an instance of the network for each previous time point, and connect these. Normally, we set a maximum number $n$ of time points we consider in history, where $n$ is significantly smaller than the number of instances $N$. The algorithm is naturally called backpropagation through time (see e.g. [126] for a more extensive discussion).

An example is shown in Fig. 8.13 where we move $n$ steps back in time. This means that we end up with a network consisting of $n$ subnetworks representing one point in time. This combined network is *without* cycles. We also do this for cases where not only the output is fed back as input, but other recurrent connections exist. This trick allows us to use backpropagation again! Of course, we do have to include the notion of time that we have now added. The precise update rule is expressed below. Note that we use brackets behind some variables to express time explicitly to stay in line with our previous explanation of the backpropagation algorithm. Furthermore, similar to the backpropagation explanation we use brackets in superscript to identify the neuron for $\hat{y}$.

For weights of non-recurrent connections:

$$\Delta w_{ij} = \eta \delta_j(t) \hat{y}_t^{(i)} \tag{8.18}$$

$$\delta_j(t) = \begin{cases} \varphi'(v_j(t))(y_t^j - \hat{y}_t^{(j)}) & \text{if } j \text{ is an output node} \\ \varphi'(v_j(t)) \sum_1^k (\delta_k(t) w_{jk}) & \text{otherwise} \end{cases} \tag{8.19}$$

For weights of recurrent connections:

$$\Delta w_{ij} = \eta \delta_j(t) \hat{y}_{t-1}^{(i)} \tag{8.20}$$

$$\delta_j(t-1) = \varphi'(v_j(t-1)) \sum_1^k (\delta_k(t) w_{jk}) \tag{8.21}$$

We see that the update rule remains the same for the non recurrent connections, although we did explicitly add the time factor to it. For the recurrent connections, we use a slightly modified version, since we are shifting between networks representing two different time points. The update of the weight between neuron $i$ (being the output of a network at the previous time step $t-1$) and a node $j$ (being a node in the network for the time point $t$) is defined similar to what we have seen earlier. We do, however, take the calculated output value at the previous time point $\hat{y}_{t-1}^{(i)}$ and the delta of the connecting node $j$ at time point $t$: $\delta_j(t)$. $\delta_j(t-1)$ propagates the error back to the network for time $t-1$ from node $j$ connected to the network for time $t$.

As you can see, we do not explicitly represent the weights by means of time. However, for each network for a time point we do derive an update for the weights. We will simply sum all updates to derive the new (universal) weight across all connected networks.

The type of recurrent neural networks we have just explained suffers from some problems. For example, they have problems representing a long term memory. Variants have been developed that tackle these issues, and are commonly used in practice. Examples are LSTM [60] and GRU [37], both have recently been applied success-fully for example in speech recognition and text analysis. Other application areas which use data similar to the quantified self context include recommender systems and user behavior analysis [77]. It is beyond the scope of this book to discuss these algorithms in detail. We will, however, focus on an alternative approach that takes a completely different perspective.

### 8.3.2    Echo State Networks

Although backpropagation through time is quite elegant, it does come at a price: high computational overhead, and the probability of getting stuck in a local minimum. More recent developments for learning of temporal patterns using neural networks include *echo state networks* [63] as part of field of *reservoir computing*. The principle of reservoir computing is to have a huge reservoir of fully connected hidden neurons with randomly assigned weights. The weights from the input layer to the hidden neurons (that are again fully connected) are also randomly generated. The only part that is not set randomly are the weights from the hidden neurons to the output layer. The connections in the reservoir can be cyclic in the case of echo state networks, thus allowing for the representation of temporal patterns. The non-cyclic case is called *extreme learning machines*. Let us look at the approach in a bit more detail. Figure 8.14 shows an example of an echo state network. Note that we did not draw all connections from the input to the reservoir and from the reservoir to the output.

Let us specify the network in a formal way. $p$ is the number of input units while we consider $n$ internal neurons and $l$ output units. Besides $x_i$ (our $i$th training vector) we specify $r_i$ as the vector with the activations of all neurons in the reservoir and $\hat{y}_i$ as the output values. The weights of the neural networks are expressed by means of matrices:

- $\mathbf{W}^{\text{in}}$ is an $n \times p$ matrix for the weights from the input layer to the reservoir.
- $\mathbf{W}$ is the $n \times n$ matrix of the internal weights in the reservoir.
- $\mathbf{W}^{\text{out}}$ is the $l \times n$ matrix that specifies the weights to the connections between the reservoir and the output.

Sometimes an additional set of connections going back from the output into the reservoir is considered but we do not take that into account here. As said, $\mathbf{W}^{\text{in}}$ and

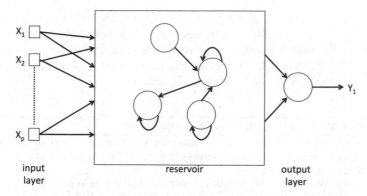

**Fig. 8.14** Echo state network (note that the input layer is fully connected to all neurons in the reservoir and the same holds for the reservoir with the output layer)

**W** are randomly initialized and are fixed, while **W**$^{\text{out}}$ is learned. We can see that the activation of the network can be calculated as follows (assuming an activation function $\varphi$ and $\varphi_{out}$ for the reservoir and the output layer respectively):

$$r_{i+1} = \varphi(\mathbf{W}^{\text{in}}x_{i+1} + \mathbf{W}r_i) \tag{8.22}$$

$$\hat{y}_{i+1} = \varphi_{out}(\mathbf{W}^{\text{out}}r_{i+1}) \tag{8.23}$$

Ok, so how do we learn? Well, just do the same that we have done before. We can feed our temporal training data in a time ordered fashion and find a weight matrix **W**$^{\text{out}}$ that minimizes the error between the desired output **Y** and the predicted output over the entire time series. Usually, the pseudo inverse method is used in the case of echo state networks. We do specify an initialization period which we do not use in our training nor in the evaluation of the generalizability upon the test set (the network needs to stabilize first), this is called the *washout time*.

Nice, simple, and lean—but does it work well? In fact, it does. It has been shown to outperform other regression and classification algorithms in a variety of cases. Furthermore, the computational properties are very nice, and the network is very expressive (in fact, as expressive as regular neural networks). So where is the catch? There is a catch when it comes to the initialization of the reservoir. It needs to be initialized such that the echo state property is satisfied (cf. [62]):

**Definition 8.1** Echo state property: The effect of a previous state $r_i$ and a previous input $x_i$ on a future state $r_{i+k}$ should vanish gradually as time passes (i.e. $k \rightarrow \infty$) and not persist or even get amplified.

In other words the reservoir should never amplify or let values of states/neurons fully persist. Otherwise, it will be impossible to learn the temporal patterns we envision. Unfortunately no formal procedures exist that are guaranteed to give us a reservoir with the echo state property. However, a heuristic exists that works well in practice. If you are unfamiliar with the terminology that follows there is no need to worry, just skip the rest of this paragraph then. The initialization considers the *eigen vectors* of the matrix $\mathbf{W}$. Once the highest eigen value (also called spectral radius, noted as $\rho(\mathbf{W})$) associated with these vectors is such that $\rho(\mathbf{W}) < 1$ we almost always see that the network satisfies the echo state property. Algorithm 19 (cf. [62]) shows how the matrix $\mathbf{W}$ can be initialized in this way.

---

**Algorithm 19:** Reservoir initialization procedure

1. Randomly initialize an internal weight matrix $\mathbf{W_0}$. $\mathbf{W_0}$ should be sparse and have a mean of 0. The size $n$ reflects the number of training examples $N$ (should not exceed $\frac{N}{10}$ to $\frac{N}{2}$ depending on the complexity)
2. Normalize $\mathbf{W_0}$ to matrix $\mathbf{W_1}$ with unit spectral radius by putting $\mathbf{W_1} = \frac{1}{\rho(\mathbf{W_0})}\mathbf{W_0}$
3. Scale $\mathbf{W_1}$ to $\mathbf{W} = \alpha\mathbf{W_1}$ where $\alpha < 1$, whereby $\rho(W) = \alpha$
Then $\mathbf{W}$ is a network with the echo state property ("has always found to be")

---

The lower the value of the parameter $\alpha$ the faster the dynamics of the reservoir are.

## 8.4   Dynamical Systems Models

While we have seen models with a lot of expressivity, we started without any knowledge on the precise form of the relationships between the different attributes and targets: these had to be learned from the data. What if we do have some knowledge about the form of the relationships, for instance obtained from literature in the specific domain? Obviously, we want to exploit this knowledge. An approach to build domain knowledge-based models that cover temporal relationships are so-called *dynamical systems models*. The models represent the temporal relationships between attributes and targets by means of differential equations (of arbitrary complexity) and assume only numerical states. An example is illustrated in Fig. 8.15. In the figure, the circles represent our target concepts (i.e. the $Y_i$'s) while the squares represent the attributes we want to use to make predictions (i.e. $X_1, \ldots, X_p$). These attributes are seen as external inputs that can influence our target concepts. To represent the next value of a certain target concept (e.g. for $Y_1$) we can use the observed values for the attributes $X_1, \ldots, X_p$ at the current time point as well as the current values of our targets. A directed arrow in Fig. 8.15 indicates that the value of the source of the connection is used to calculate the value of the target at the next time point.

### 8.4.1   Example Based on Bruce's Data

Let us consider an example of a model, and take the one shown in Fig. 8.1 again. We want to model the relationship between the *mood* and the *activity level* normalized to the [0, 1] domain, and consider both as a target for our model. We use one additional

**Fig. 8.15** Example
dynamical systems model

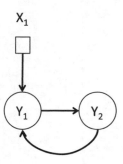

input, not being a target, namely whether the person went outside of the house during a day or not represented by 1 and 0 respectively. An example of a simple dynamical systems model that represents this problem is shown below. Note that we, for the sake of clarity, use a notation with names in subscript instead of indices in superscript to identify the variables. We use brackets to denote the time point instead of a subscript for the same reason.

$$\hat{y}_{mood}(t + \Delta t) = y_{mood}(t)+$$
$$x_{outside}(t) \cdot (\gamma_1 \cdot (1 - y_{mood}(t)) \cdot pos(y_{activity\ level}(t) - y_{mood}(t))+$$
$$\gamma_2 \cdot y_{mood}(t) \cdot neg(y_{activity\ level}(t) - y_{mood}(t))) \cdot \Delta t \qquad (8.24)$$
$$\hat{y}_{activity\ level}(t + \Delta t) = y_{activity\ level}(t)+$$
$$(\gamma_3 \cdot (1 - y_{activity\ level}(t)) \cdot pos(sin(\frac{t - \gamma_4\pi}{\gamma_5}))+$$
$$\gamma_4 \cdot y_{activity\ level}(t) \cdot neg(sin(\frac{t - \gamma_4\pi}{\gamma_5}))) \cdot \Delta t \qquad (8.25)$$

where

$$pos(v) = \begin{cases} 0 & v < 0 \\ v & otherwise \end{cases} \qquad (8.26)$$

$$neg(v) = \begin{cases} v & v < 0 \\ 0 & otherwise \end{cases} \qquad (8.27)$$

While the model might seem complex, it behaves in a reasonably straightforward way. One of the basic starting points is that we want to keep the values of the states between 0 and 1. Hence, we can at most increase a state with 1 minus its current value, and we cannot decrease a state more than its current value. If we look at the equation for $\hat{y}_{mood}$, we see that we determine the next value $\hat{y}_{mood}(t + \Delta t)$ by taking its previous value, and seeing how its value relates to the *activity level*. If the *activity level* is above the *mood*, it will increase the *mood*, otherwise it will decrease

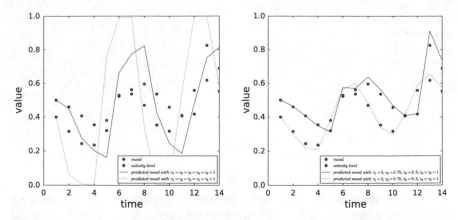

**Fig. 8.16**  Example output using two parameter settings

the *mood*. Note that this influence only holds in case of an outside activity. For the *activity level*, we see that the value increases or decreases based on a sine function. You can see that a number of parameters are present: $\gamma_1$ to $\gamma_5$. We use a fixed step size $\Delta t$ as we have seen before.

Figure 8.16 shows an example of our model with two sets of parameter settings, initialized at time point 1 with the same value (though scaled to [0, 1]) as we have seen in our example figure from Bruce in Fig. 8.1. We assume that all days have at least one outdoor activity (i.e. $X_{outdoor} = 1$ for all time points). We set $\Delta t$ to 1 and run the model for 14 time steps (representing days). This means that we predict the target values at $t + 1$ based on the predicted value at time $t$ plus the external inputs. We see that the parameter values have a huge influence on the predictions in the figure. The figure on the right is clearly much closer to the measured data. We will discuss ways to optimize the parameter values next. For some more extensive examples of dynamical system models that predict human behavior, see e.g. [23–25].

### 8.4.2  Parameter Optimization

We can use this model to make predictions for the future, even beyond a single time point. However, we would need to know the value for the external attributes (in this case $X_{outside}$) as our models depends on that. If we were to predict over multiple time points in the future without knowing the external output, we could either take an average value for this, or sample from a distribution that is in line with the previous observations of the variable.

So where does the machine learning come into play? As said, the setting of the parameters is crucial and highly dependent on the individual. Machine learning can help in finding the best settings for these parameters. Hence, we seek parameter values that minimize the difference between the actual target value and the predicted ones (i.e. $y(1), \ldots, y(N)$ versus $\hat{y}(1), \ldots, \hat{y}(N)$ in our notation with brackets). We could treat numerous approaches that find the best parameters. We will discuss two heuristics, namely *simulated annealing*(cf. [71]) and *two variants of genetic algorithms* (see e.g. [44] for a broad overview of evolutionary algorithms, including genetic algorithms). We could again use the quadratic error function of a model with parameter values $\lambda$:

$$E(\lambda) = \sum_{t=1}^{N} (\hat{y}(t) - y(t))^2 \tag{8.28}$$

where $\hat{y}(t)$ is the prediction of the model given $\lambda$. We need to find parameter settings that minimize this error function.

### 8.4.2.1  Simulated Annealing

When we consider simulated annealing we make random steps in the parameter space and see whether we make improvements in terms of performance. There is something smart to it though: moves in the parameter space that do not have a positive impact on the error are still considered with a certain probability. Doing so, helps us to explore the parameter space. As the algorithm keeps running, the probability to accept'negative' moves decreases. This depends on the notion of a *temperature*—thus the term annealing. The lower the temperature becomes the less we explore the search space. Consider the Algorithm 20. The algorithm performs a number of iterations ($k_{max}$) and in each iteration tries a new update of the parameter vector $\lambda$. When taking the random steps we need to make sure the parameter values are within the desired ranges. If the new vector performs at least as good as our current one we take that as the new current one. If it does not, we accept it with a probability depending on the current temperature and the parameter $k_b$. This probability decreases when the temperature decreases (by a factor $\alpha$ which is assumed to be less than 1). The algorithm is simple and elegant, and in general performs quite well.

### 8.4.2.2  Genetic Algorithms

An alternative to the simulated annealing scheme presented above are genetic algorithms (GA). They work based on the theory of evolution. While there have been many advancements in the field (see e.g. [44]), we will explain only the so-called *simple GA* to get a feeling for the type of approach. We follow the explanation of Eiben and Smith [44]. The basic starting point of the algorithm is a population of candidate solutions, in our case parameter vectors. In genetic algorithms these candidates are represented by means of binary strings (the so-called *genotype*). We can allocate a number of bits in the string per parameter value. The number of bits depends on the desired granularity. Assuming we use $n$ bits to represent one parameter value, we

---

**Algorithm 20:** Simulated Annealing

---
$\lambda_{current}$ = random
$E_{prev} = \infty$
Temp = Temp$_{init}$
**for** $k$ *from 1 to* $k_{max}$ **do**
    **for** $i$ *in* $\lambda$ **do**
       |  $\lambda'_i = \lambda_i +$ random
    **end**
    **if** $E(\lambda') \leq E(\lambda_{current})$ **then**
       |  $\lambda_{current} = \lambda'$
    **else if** $e^{\frac{(E(\lambda_{current})-E(\lambda'))}{k_b Temp}} \geq random(0, 1)$ **then**
       |  $\lambda_{current} = \lambda'$
    Temp $= \alpha \cdot$ Temp
**end**
**return** $\lambda_{current}$

---

**Fig. 8.17** Parameter Vector genotype

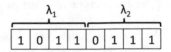

end up with a genotype of length $n \times |\lambda|$. An example of a bit string representing a parameter vector $\lambda$ with two values, each represented by 4 bits, is shown in Fig. 8.17.

We start with a certain random population of these candidates. We then go into an evolutionary process very much inspired by natural evolution: we perform selection of parents for a mating pool, perform crossover and mutation, and select a new population for the next generation. This loop continues for a set number of *generations*. Let us consider each one of these steps in a bit more detail.

**Parent selection**: First, we need to select parents for our mating pool. This is done by means of the assignment of a *fitness value* to each of our individuals. In our case this would be the error of the parameter vector on the training set (cf. Formula 8.28). We select using a so-called *roulette wheel* technique: we assign probabilities to individuals being selected based on their fitness. We select the same number of individuals as the population size. We assign the probability of an individual $i$ (which represents a parameter vector $\lambda_i$) being selected for a single spin of the wheel as follows:

$$P_i = 1 - \frac{(E(\lambda_i))}{\sum_{j=1}^{pop\_size} E(\lambda_j)} \tag{8.29}$$

**Crossover:** Once we have selected the individuals into the mating pool, we select pairs of individuals (without using any of them twice) and perform crossover with a probability $p_c$. Crossover for the simple GA works with a single point: you randomly select a point in the bit string and create two children. For child one we take the first part (before the crossover point) of parent one and the second part (after the

**Fig. 8.18** Simple GA single
point crossover

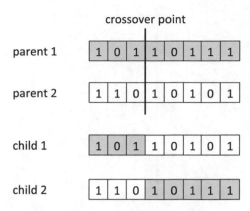

crossover point) from the second parent. For the second child we do the opposite.
The process is illustrated in Fig. 8.18. If we do not perform crossover (since $p_c$ will
be lower than 1) we take the two original parents.

**Mutation:** Finally, we perform mutation with a certain probability $p_m$. We do this
on each of the individuals resulting from the previous step. It works by flipping bits
in the bit strings of individuals with probability $p_m$. The final resulting individuals
become the new population.

A brief summary of the evolutionary loop for the simple GA is shown in Algorithm 21.

---

**Algorithm 21:** Simple Generic Algorithm

---

population = random initialization of population with set population size ps
**for** *i from 1 to max_generations* **do**
    Select ps parents according to Eq. 8.29
    Select pairs of parents from the individuals we have selected (without replacement)
    Apply crossover to the parents with probability $p_c$ or copy the original parents
    Apply bitwise mutation with probability $p_m$ on the resulting individuals
    The individuals we have just created become the new population
**end**
**return** *fittest individual in the final population*

---

### 8.4.2.3  Multi-criteria Optimization

The dynamical systems models describe the relationship between multiple target
concepts. Often, trade-offs need to be made in the setting of the parameters: adjusting
a parameter might improve the predictions on one target, while it worsens them for
another target. If we assume all are equally important, we can just optimize the mean
squared error between the target values using the methods we listed before. If we

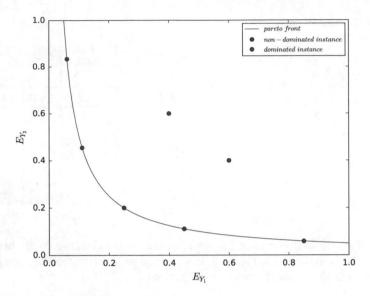

**Fig. 8.19** Error score model instances with Pareto Front

want to study the trade-off between the performance on each of the targets we have a *multi-criteria optimization problem.*

If we return to our previous example, we could end up with a series of parameters settings for $\gamma_1$ to $\gamma_5$, where each setting results in an error on the *mood* target and an error on the *activity level* target. We call a model with these parameter settings a *model instance.* Hence, we have separate error functions per target:

$$E_{target}(\lambda) = \sum_{t=1}^{N} (\hat{y}_{target}(t) - y_{target}(t))^2 \tag{8.30}$$

We denote the error of a specific model instance with parameter vector $\lambda$ for a target $i$ as $E_i(\lambda)$. We assume $q$ targets. Let us consider Fig. 8.19. The dots represent model instances with their mean squared errors for our two targets. We can see that some instances score a lower error on the *mood* while others score better on *activity level.* Some model instances clearly do not have a preferable parameter setting, because other models score better on one of the two targets without performing worse on the other. These are depicted in blue. Of course, these are not the model instances we would choose. Instead, we only want to find model instances that are not, as we say, dominated by other models.

**Definition 8.2** A model instance $\lambda_m$ is dominated by another model instance $\lambda_n$ when the mean squared error obtained by model instance $\lambda_n$ is lower for at least one target and not higher for any of the other targets.

More formally:

$$dominated(\lambda_m, \lambda_n) = \begin{cases} 1 & \exists i \in 1, \ldots, q : E_i(\lambda_m) > E_i(\lambda_n) \wedge \\ & \forall j \in 1, \ldots, q : E_j(\lambda_m) \geq E_j(\lambda_m) \\ 0 & \text{otherwise} \end{cases} \qquad (8.31)$$

In the end, we are interested in finding a nice set of non-dominated model instances. These non-dominated model instances make up the so-called *Pareto Front* which connects the model instances, shown as a red line in Fig. 8.19. In case of more targets we can go beyond a two-dimensional representation. Once the model instances are found, a domain expert can select a model deemed appropriate given the importance of targets in the domain. For finding these model instances that are positioned on the Pareto Front we have a number of options at our disposal. One is the well-known algorithm NSGA-II (for Non-Dominated Sorting Genetic Algorithm-II) [40].

We have explained the concept of an evolutionary algorithm, namely the simple GA, earlier. NSGA-II works more or less based on the same principles in terms of the representation (bit string again), and the mutation and crossover operators. Of course our goal is now different: we want to find a good spread of points that are positioned on a Pareto Front, we no longer have a "best" solution as we are considering the trade-offs between the predictions on multiple targets. Let us first consider a population $P$ of model instances. We first need to identify sets of points that together form a Pareto front of non-dominated solutions, expressed in Algorithm 22. This algorithm forms multiple Pareto Fronts of non-dominated solutions, identified as $F_i$. We start with the first model instance in the population and take that as a basis for a Pareto Front. We add model instances to the front and remove them when they are dominated by other model instances in the front. After we have passed all model instances in our population, we have our first front (so we made a selection of our model instances), although we do not know how optimal it is. We then continue with the remaining model instances that were not included in the front yet and look for a next collection of these remaining model instances that constitute a next front, etcetera.

We now have a set of $f$ Pareto Fronts $F_1, \ldots, F_f$. Another aspect is how well solutions are spread across our Paterto Front. We would like to have a collection of solutions that are nicely spread across the front otherwise we are not quite sure on the shape of the front, and the expert will not be able to select a model that represents specific trade-offs. To establish this, we first need to determine the distance between points in our fronts. This is performed using calculation expressed in Algorithm 23. In the algorithm, we see that model instances of a specific Pareto Front $F_i$ are initialized to have a zero distance. Then, for each objective they are sorted based on their error for that objective. Points that score highest or lowest on the objective are set to an infinite distance as they are the most extreme points on the front that we want to keep for sure. This is because we want have the maximum width of our front and well will see that bigger distance means a higher chance of being selected. Distance between the other points is the difference between the error for the two adjacent points.

---

**Algorithm 22:** Finding Pareto Fronts

---

**find_pareto_fronts(P):**
used = []
i = 0
F = []
**while** $|P| > 0$ **do**
    $P' = P[1]$ // The first model instance in the population
    **for** $p \in P \wedge p \notin P'$ **do**
        $P' = P' \cup \{p\}$
        **for** $q \in P' \wedge p \neq q$ **do**
            **if** *dominated(q, p)* **then**
                | $P' = P' \setminus \{q\}$
            **else if** *dominated(p, q)* **then**
                | $P' = P' \setminus \{p\}$
        **end**
    **end**
    Add $P'$ to $F$
    $P = P \setminus P'$
    $i = i + 1$
**end**
**return** $F$

---

**Algorithm 23:** Finding distances between points on the Pareto Front

---

**distance_assignment($F_i$):**
$l = |F_i|$
distance = []
**for** $j$ in $1, \ldots, l$ **do**
    | distance[j] = 0 // Initialize the distance to zero
**end**
**for** $k$ in $1, \ldots, q$ **do**
    $F_i = sort(F_i, k)$ // Sort the model instances based on their error for objective $k$
    distance[$F_i[1]$] = distance[$F_i[l]$] = $\infty$ // Make sure boundary points are always selected
    **for** $s$ in $2, \ldots, (l - 1)$ **do**
        | distance[$F_i[s]$] = $E_{F_i[s],k} + (E_{F_i[s+1],k} - E_{F_i[s-1],k})$
    **end**
**end**
**return** *distance*

---

We are finally ready to look at the overall algorithm (Algorithm 24). This shows how we proceed from one generation to the next. The initialization can be done in a random way. In the algorithm, a combined population of the parents and children is created. From these we identify the Pareto fronts following the algorithm we have previously seen. We then create a new population of parents by starting with the first (and most dominant) Pareto Front we have identified, and add the complete set of

---

**Algorithm 24:** NSGA-II main loop

---

$R_t = P_t \cup C_t$ // Take the parent and child population
$F_1, \ldots, F_f = \text{find\_pareto\_fronts}(R_t)$
$P_t = \emptyset$
$i = 1$
**while** $|P_{t+1}| < |P_t|$ **do**
    **if** $|P_{t+1}| - |P_t| \geq |F_i|$ **then**
        $P_{t+1} = P_{t+1} \cup F_i$
        $i = i + 1$
    **else**
        $d = \text{distance\_assignment}(F_i)$
        $F_{sorted} = \text{sort}(F_i, d)$
        $P_{t+1} = P_{t+1} \cup \{F_{sorted}[1]\} \cup \cdots \cup \{F_{sorted}[|P_{t+1}| - |P_t|]\}$
**end**
Create $C_{t+1}$ using crossover and mutation
$t = t + 1$

---

model instances that span up the front. We continue until we can no longer add all the model instances of the current Pareto Front due to the limit on the number of parents. In that case, we add the instances that are most apart from the other instances (using the distance function defined before) in the front. Once we have the set of instances, we generate the offspring by applying crossover and mutation and move to the next step. In the end, this gives us a nice Pareto Front which we can use to select the most suitable model instance.

This ends our part on dynamical systems models. Note that more complex functions to define how appropriate a model is have been defined as well, see e.g. [120].

## 8.5 Case Study

We return to the crowdsignals dataset. We focus on the temporal regression problem to predict the heart rate that we identified earlier and will apply several approaches we have introduced in this chapter. The selection of the data and features do not change if we apply the temporal learning algorithms, therefore we will not touch upon these aspects again.

### 8.5.1 Tuning Parameters

Previously, we tuned the parameters by means of a cross validation approach on the training set and identified the parameter values that performed best in the cross validation. The cross validation approach does, however, not take the order of the data into account. Hence, we use a different scheme for our temporal learning algorithms as they need to be trained on temporal sequences of data. We use a fixed (connected)

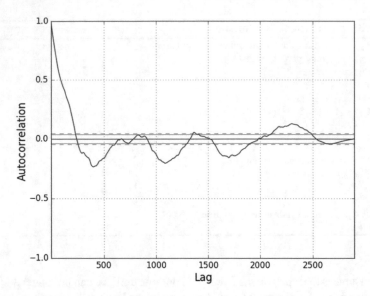

**Fig. 8.20** Autocorrelation for different lags for *hr_watch_rate*

interval within the training set to train our models with certain parameter values and
test their performance on the remaining interval. We select the parameter values that
perform best and then train on the entire training set. We made the split as shown by
the dashed line in Fig. 7.17 as this split contains a nice coverage of activities in both
parts of the split. The tuning parameters of the algorithms are shown in Table 8.1.
Furthermore, we normalize the data (to a [0.1, 0.9] range for the Recurrent Neural
Network and [−0.9, 0.9] for the Echo State Network) to boost performance. Finally,
we translate the predicted values back to the original range.

For the time series we first of all check whether the heart rate is stationary and
we explore the autocorrelations in the series. Using the Dickey-Fuller test we find
that the series is indeed stationary (with a p-value of 0.016). For the autocorrelations
with different lags we show a visualization in Fig. 8.20. We see that lower lags result
in a high autocorrelation. This means that we will try low values for the lag in
our parameter tuning phase. Furthermore, we use a variant of ARIMA which also
considers latent or exogenous variables (i.e. features from other sensors) and assigns
weights to them. Given the limits of the algorithms associated with time series (they
cannot cope with lots of features) we only apply the times series to our initial set of
measurements from Chap. 2.

**Table 8.1**  Algorithms and parameters for the regression problem

| Algorithm | Variant description | Parameters varied |
|---|---|---|
| Echo State Network (ESN) | Randomly connected reservoir of neurons with tanh activation function with the output being fed back into the reservoir | Number of neurons in reservoir: {400, 700, 1000} $\alpha$: {0.6, 0.8} |
| Recurrent Neural Network (RNN) | Recurrent neural network with one layer of hidden neurons with a sigmoid activation function and sigmoid output nodes | Number of hidden neurons: {50, 100} maximum iterations over the entire dataset: {250, 500} |
| Time series | ARIMAX algorithm using Bayesian inference | p: {0, 1, 3, 5} q: {0, 1, 3, 5} d: {0, 1} |

## 8.5.2  Results

The results are shown in Table 8.2. We see that the results are much worse than the results we obtained in the previous chapter, except for the time series. The echo state network seems to be able to perform reasonably well on the training set but it generalizes very bad. The recurrent neural network does precisely the opposite, which is not what we would expect to see. Possibly the parameter values have not been selected properly due to the characteristics of the validation set and the model obtained a bad fit on the full training set with these parameter settings. The time series approach performs best, but this is mainly because it just predicts approximately the average. Our hypothesis is that the data for the heart rate is too noisy and inconsistent to really identify trends over time given the amount of data. This does not hold for all measurements as we previously saw the activity is very well learnable. Furthermore, as points are treated as a sequence and the next prediction depends on the previous prediction as well, errors can easily propagate over time while this is not the case for the previous approaches discussed in Chap. 7. When we consider the subset of predictors and their relative performance we see a scattered picture but we do see that the temporal features have less of an impact on the performance compared to the approaches that do not consider the temporal dimension (and this makes sense of course).

Let us study the performance of the models a bit more closely. We take the recurrent neural network with the initial features first as this shows the best generalizability. Figure 8.22 shows the predictions (with a setting of 50 hidden neurons, and 250 epochs). We observe similar problems as we have seen in our previous chapter on top of the fact that predictions generally seem to underestimate the heart rate. Furthermore, we see pretty extreme mistakes in both the training and test set predictions, and several periods where the prediction is off for a prolonged number of time points. When we explore the echo state network with all possible features (shown in Fig. 8.23) with a reservoir size of 400 and $\alpha$ of 0.8 we observe that, except for a

**Table 8.2** Performance of algorithm on label regression task (ESN = Echo State Network, RNN = Recurrent Neural Network, TS = Time Series)

| Approach Features | ESN | | RNN | | TS | |
|---|---|---|---|---|---|---|
| | Training | Test | Training | Test | Training | Test |
| Initial set | 3107.0 (3529.1) | 4654.3 (4544.7) | 4004.8 (4558.3) | 2507.1 (2899.4) | 1567.0 (1210.6) | 1347.8 (943.8) |
| Chap. 3 | 3061.7 (3687.7) | 4609.4 (4481.7) | 3954.1 (4522.1) | 2728.5 (3058.6) | - | - |
| Chap. 4 | 3127.6 (3661.9) | 5425.9 (4926.3) | 4109.8 (4568.9) | 2877.4 (3382.9) | - | - |
| Chap. 5 | 2957.2 (3574.8) | 5603.6 (4979.0) | 4105.7 (4559.3) | 3138.3 (3618.3) | - | - |
| Selected features | 2782.3 (3253.0) | 4394.4 (4368.4) | 3774.1 (4471.3) | 2925.6 (3391.0) | - | - |

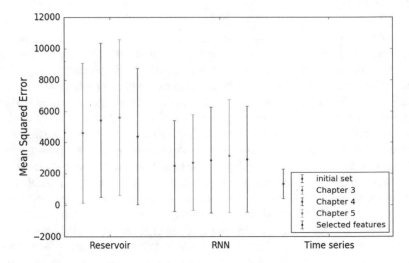

**Fig. 8.21** Performance of different algorithms

few extreme deviations, it is able to capture the training data pretty well. However, we see an extremely noisy set of predictions for the test set. The output of the time series model is shown in Fig. 8.24, with a setting $p = 0$ and $q = 5$. We see that it more or less predicts the average, so it is not doing a good job, though it does better in terms of performance than the neural network based models. Table 8.3 shows the importance of the key features used by the model.

We do not feel these results generally hold for the techniques we have explained in this chapter, they should work in case of a more extensive history. It does show that careful evaluation and a range of algorithms (both temporal and non temporal) should be explored to select the most suitable type of model for a given dataset.

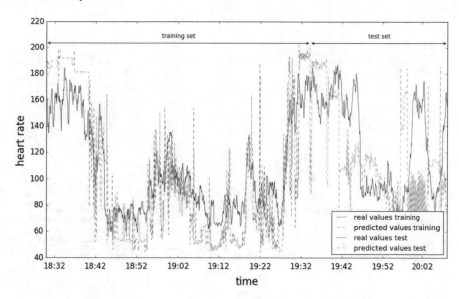

**Fig. 8.22** Actual versus predicted values for RNN

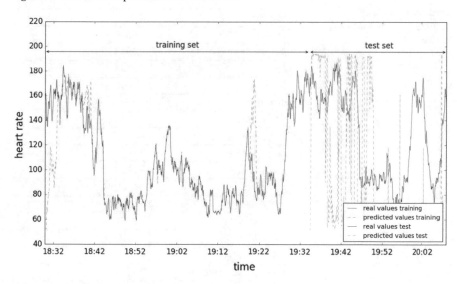

**Fig. 8.23** Actual versus predicted values for ESN

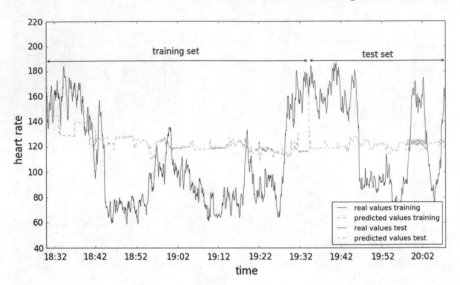

**Fig. 8.24** Actual versus predicted values for TS

**Table 8.3** Attribute importance for ARIMAX model

| Feature | $\beta$ |
|---|---|
| *MA(1)* | 1.8606 |
| *MA(2)* | 2.2943 |
| *MA(3)* | 2.0588 |
| *MA(4)* | 1.2991 |
| *MA(5)* | 0.4741 |
| *acc_phone_x* | −0.2286 |
| *acc_phone_y* | −0.3136 |
| *acc_phone_z* | 0.1977 |
| *acc_watch_x* | 0.1259 |
| *gyr_phone_y* | −0.1105 |
| *gyr_phone_z* | −0.1282 |
| *labelOnTable* | 0.2781 |
| *labelSitting* | −0.1099 |
| *mag_phone_y* | −0.1317 |
| *press_phone_pressure* | 0.1173 |

# 8.6  Exercises

## 8.6.1  Pen and Paper

1. When discussing time series analysis we defined the *acf* function. Can we have two different models that have the same *acf*? If so, give an example.
2. Recurrent neural networks are known to have a problem to represent long term memory. Approaches have been proposed that tackle this problem. Select one of these algorithms and explain how it works.
3. Explain how the training time of a recurrent neural network relates to that of a regular neural network without recurrence.
4. It seems strange that a random reservoir for echo state networks can do a similar job as we see for other neural networks while saving training time. Try to find a form of a proof that echo state networks are indeed able to learn patterns in a similar fashion and write down the highlights of the proof.
5. A lot of effort has gone into trying to find better strategies to initialize the random reservoir. However, due to the "no free lunch theorem" there is not a single approach across all datasets. Explain the concept of the "no free lunch theorem" in this context.
6. Give an example of a setting for the quantified self where you would expect a recurrent neural network to work better than a feed forward network.
7. Although we have mainly treated pure machine learning based approaches, a dynamical systems model can also be a good starting point for predictive modeling. Give two advantages of using such a model and two disadvantages.
8. For the algorithms we have discussed to tune the parameters of the dynamical systems model, explain whether they are guaranteed to find the optimal parameter setting.
9. How could we avoid overfitting towards the data for the parameter tuning of the dynamical systems model?

## 8.6.2  Coding

1. In Eq. 8.5 we demonstrate how second-order differencing can be applied to a time series. Generate a non-stationary time series as the one in Fig. 8.2c and apply second-order differencing. Describe the results.
2. Apply the neural network based approaches that have been treated here to one of your own dataset similarly to how we have done it for the crowdsignals data.

3. Develop a simple dynamical systems model to predict the heart rate based on certain features of the crowdsignals dataset (possibly with more high level feature such as "activity level" that you might need to derive from your data). Apply it to the crowdsignals dataset by tuning its parameters and discuss the predictive performance.
4. Implement two alternative strategies to initialize the reservoir in echo state networks and compare the performance to the performance achieved using the basic implementation we have provided together with this book.

# Chapter 9
# Reinforcement Learning to Provide Feedback and Support

In the approaches we have discussed so far we aimed at gaining insights from the data: we filtered out noise from data, identified new attributes, and created predictions for unseen data. This gives us a good understanding of the current state of our quantified self enthusiasts and the expected future state. However, we do not yet use our understanding to "close the loop", namely to provide feedback and possible support actions to the users given the understanding we have. The importance of this is also stressed in [76, 80, 113]. Imagine Arnold: we predict that his progress is going to stagnate if he continues his current schedule. We should actually do something with this information right? We could provide a form of feedback or advice to Arnold about how to avoid this stagnation. When we consider Bruce there is a possibility that we predict a sharp decrease in his mood over the next couple of days. If we know what actions work well to boost his mood we can advice Bruce to perform one of those actions. In this chapter we will tackle this problem of learning when to perform what actions from a reinforcement learning perspective. While applications of reinforcement learning in this domain are not widespread, the fit with our purpose is evident. We will just explain the basics in this chapter. There have been substantial advancements in the area of reinforcement learning and we refer the reader to [127] for a nice overview of the state-of-the-art combined with [79] for in-depth discussion of deep reinforcement learning. We follow Sutton and Barto [112] by sketching the basic setting and notation first (which is a bit different from our previous machine learning notations), followed by a discussion of appropriate techniques.

## 9.1 Basic Setting

Figure 9.1 shows illustrates our problem. We identify two actors in the setting, namely the *user* (normally called the *environment*, but given our setting we feel this name is more appropriate) and the *agent*. The user is one of the quantified selves, while the

© Springer International Publishing AG 2018  203
M. Hoogendoorn and B. Funk, *Machine Learning for the Quantified Self*,
Cognitive Systems Monographs 35, https://doi.org/10.1007/978-3-319-66308-1_9

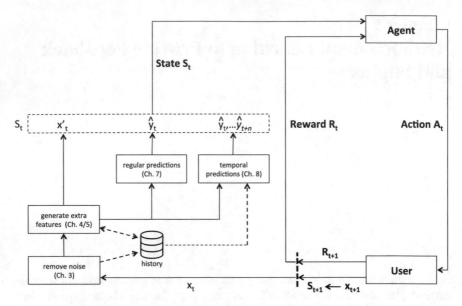

**Fig. 9.1** Our quantified self approach as a Reinforcement Learning Problem (RL loop from [112])

agent is a software entity that decides when and how to support a user (e.g. the app running on the mobile phone). The agent can observe the *state* of the user at a certain time point $t$, denoted as $S_t \in S$. Connecting to the previous chapters we make the definition of this state a bit more precise. We obtain a data instance $x_t$ from the user at time $t$ that represents the sensor values. We then apply the techniques we have previously discussed to create a representation of the state of the user based on this sensory data. This includes:

- Removal of noise (cf. Chap. 3).
- Enrichment of our original set of features without noise by additional ones (following Chap. 4 and the presence in a certain cluster as explained in Chap. 5), resulting in an enriched set $x'_t$
- Prediction of unknown values using the traditional machine learning techniques explained in Chap. 7: $\hat{y}_t$
- Prediction of unknown (future) values using the machine learning techniques explained in Chap. 8, resulting in $\hat{y}_t, \ldots, \hat{y}_{t+n}$ if $n$ time points in the future are predicted. Note that making predictions for the future can be tricky as we now intend to influence the future by intervening. In reinforcement learning there are specific ways in which this is incorporated. We will return to this discussion later.

For some approaches we might require a history of measurements (e.g. temporal features and predictions), indicated by the database and dashed arrows in the figure. Together these values form our state $S_t$. Based on $S_t$ the agent can make a decision on a certain *action* $A_t$, taken from the set of available actions in the current state, i.e.

$A_t \in \mathcal{A}(S_t)$. The applied action results in a new state of the user ($S_{t+1}$ obtained via processing $x_{t+1}$) and a so-called *reward* $R_{t+1}$. The reward is the goal in reinforcement learning problems. It can be defined based on the state of the user and maps it to a numerical value. For instance, the overall state of happiness of Bruce at a certain time point. We are not only interested in the next reward we obtain, but in the total rewards we expect to accumulate in the future, called a *value function*. An action to support Bruce might for example not have an immediate effect while it does turn out to be best in terms of future rewards. In reinforcement learning we try to find what is called a *policy*, expressing which action to perform in what state. Policies are deemed suitable when the application of the policy results in a lot of rewards over time (i.e. we perform well in terms of the value function). Approaches should strike a good balance between *exploration* (trying out new actions to see how successful they are) and *exploitation* (applying known successful actions). If we perform too little exploration we might never perform a potentially very successful action while if we never exploit we could end up trying actions that are mostly not appropriate.

Let us formulate the problem we are facing a bit more precise and formal (yes, expect a good number of formulae to follow shortly). We define the goal $G_t$ at time point $t$ by means of the future reward we obtain:

$$G_t = R_{t+1} + \gamma R_{t+2} + \gamma^2 R_{t+3} + \cdots = \sum_{k=0}^{T-(t+1)} \gamma^k R_{t+k+1} \tag{9.1}$$

Here, $\gamma$ is a *discount factor* for the future reward in the range $[0, 1]$. It indicates the relevance of these for the evaluation. In case of a value $\gamma = 0$ we only care about the immediate reward while for a value of $\gamma = 1$ future rewards are equally important as the current reward. $T$ expresses the end time we consider in our system (which can also be $\infty$). One important aspect in several approaches we will discuss is that the system we described in Fig. 9.1 has the *Markov Property*. To define this property in a formal way, let us consider our states and rewards. We can express a certain probability that we end up in a state $S_{t+1} = s$ with a reward $R_{t+1} = r$ based on our entire history:

$$\mathbf{Pr}\{R_{t+1} = r, S_{t+1} = s | S_0, A_0, R_0, \ldots, S_t, A_t, R_t\} \tag{9.2}$$

We can also only consider the history of the previous time point without considering the reward at the time point:

$$\mathbf{Pr}\{R_{t+1} = r, S_{t+1} = s | S_t, A_t\} \tag{9.3}$$

We say that the state property has the *Markov Property* when both probabilities are equal for all rewards $r$ and states $s$ over all time points. In other words, when the previous state and action contain enough information to assess the probabilities of moving to a next state properly. If we think of a game of chess, this property is satisfied: if we know the position of the different pieces on the board and the movement performed as that is all information that is relevant for the game, the entire history is not. Is the property likely to hold for the quantified self? Good question, which highly depends on the amount of sensory data obtained and the problem we aim to tackle. It is very likely that it will not be satisfied for quite some cases. A lot of approaches do make this assumption though, and they often consider the problem as a finite Markov decision process (finite MDP) with the hefty assumption of a finite number of states and actions. Given our input space (where we typically have lots of numerical values) we cannot expect this for our case, but there are ways around this, namely to map our space to finite domains. We will turn to this later.

Assuming our problem to be a finite MDP though, we define the probability of moving to a next state $s'$ from a state $s$ with action $a$, referred to as the *transition probability*:

$$p(s'|s, a) = \mathbf{Pr}\{S_{t+1} = s'|S_t = s, A_t = a\} \tag{9.4}$$

We express the expected reward:

$$r(s, a, s') = \mathbb{E}[R_{t+1}|S_{t+1} = s', S_t = s, A_t = a] \tag{9.5}$$

This formulates our problem in a very nice way. To give a concrete example, of $p(s'|s, a)$ imagine Bruce in a depressed state, we provide Bruce with a message "cheer up" and predict the probability that Bruce will no longer be depressed at the next time step. Furthermore, we calculate the expected reward $r(s, a, s')$ for that outcome, which will be high as we move to a non-depressed state.

The final aspect we need to fully formulate our problem is the policy $\pi$, i.e. when to perform what action. Each policy assigns a probability to a certain available action $a$ in a given state $s$: $\pi(a|s)$. Given the assumptions we have previously made, the expected *value* of a state $s$ (state-value function) if we would follow our policy $\pi$ thereafter is:

$$v_\pi(s) = \mathbb{E}_\pi[G_t|S_t = s] \tag{9.6}$$

And similarly, we define the value of an action $a$ in state $s$ given our policy $\pi$ as (called the action-value function):

$$q_\pi(s, a) = \mathbb{E}_\pi[G_t|S_t = s, A_t = a] \tag{9.7}$$

We are interested in finding the policy (or policies) $\pi_*$ that provides the highest state-value function in all states:

$$\forall s : v_{\pi_*}(s) \geq v_\pi(s) \tag{9.8}$$

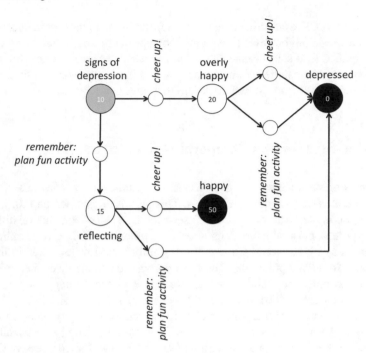

**Fig. 9.2**  Example reinforcement learning Problem for Bruce

The value is denoted as $v_*(s)$. Similarly, we define the *optimal action value function*: $q_*(s, a)$ given our optimal policies $\pi_*$. Well, the stage is set. We have treated all the fundamentals.

Let us consider a simple example to understand matters a bit better. Figure 9.2 shows such an example. We assume that the state resulting from an action is completely deterministic in this case. The figure shows an overview of the relevant states of Bruce (i.e. $S$) related to his depression problem. They are presented by the large circles. We have an initial state (*signs of depression*), two intermediate states (*overly happy* and *reflecting*) and two terminal states: *depressed* and *happy*. Obviously we do not mean Bruce will cease to exist when he falls into a depression, but given his history he will likely be advised to seek human counseling which is beyond the scope of our system. In a happy state there is no need to support Bruce any longer until our friend shows signs of depression yet again. In each state, we have the same two actions at our disposal ($\mathcal{A}(S)$ is the same for all $S \in \mathcal{S}$). The first action asks Bruce to *remember to plan fun activities*, something he should recollect from prior therapies. The second option is to give Bruce a boost by providing him the message *cheep up!*. The figure expresses the transitions between states based on these actions by means of arrows with a small circle representing the action, e.g. when we perform the action *cheer up!* in the *signs of depression* state, Bruce moves to the *overly happy* state. Each state also shows the reward it gives. What would be the optimal policy for our initial state in this case? Well, assuming that we only consider the reward of the next state it would be selecting *cheer up!* resulting in the highest immediate reward of 20. In the long run however (if we only look ahead one additional time point) the obvious choice is to select *remember: plan fun activity* as we can get Bruce in the *happy* state after one more action then and we obtain a high reward for that. How do we find such a policy $\pi$? We will look into that now.

## 9.2   One-Step SARSA Temporal Difference Learning

In the first approach we are going to discuss we assume that we can make a table of the situations versus the possible actions. This algorithm is based on [102]. We do not assume complete knowledge on the user, making a so-called temporal difference based approach most suitable. This allows for online learning (i.e. learning from experiences as we go along) while still using information we obtained in the past. Our goal is to learn a policy that gives us appropriate actions given a certain state, also called a control problem. An alternative is to learn the valuing of a state (which we will not consider in the remainder of this chapter). Let $Q(S_t, A_t)$ denote the value of the *state-action* pair for a policy $\pi$. This represents how appropriate action $A_t$ is in situation $S_t$ given our policy $\pi$, i.e. how suitable a transition is between states. For instance, how appropriate the action to say "cheer up" for a depressed Bruce is. Remember that we have previously defined $q_\pi(s, a)$. We can alter this definition a bit, by saying that instead of looking at the goal $G(t)$ (i.e. the future rewards), we look at the reward we obtain in the next state and the state-action value of our policy given the next state. We derive a new definition as follows:

$$q_\pi(s, a) = \mathbb{E}_\pi[G_t | S_t = s, A_t = a] \tag{9.9}$$

$$q_\pi(s, a) = \mathbb{E}_\pi[\sum_{k=0}^{T-(t+1)} \gamma^k R_{t+k+1} | S_t = s, A_t = a] \tag{9.10}$$

$$q_\pi(s, a) = \mathbb{E}_\pi[R_{t+1} + \gamma \sum_{k=0}^{T-(t+2)} \gamma^k R_{t+k+2} | S_t = s, A_t = a] \tag{9.11}$$

$$q_\pi(s, a) = \mathbb{E}_\pi[R_{t+1} + \lambda q_\pi(S_{t+1}, A_{t+1}) | S_t = s, A_t = a] \tag{9.12}$$

We want to create an estimate of the value of an action $A_t$ in a situation $S_t$ given our policy $\pi$. We define this estimate as $Q(S_t, A_t)$. Based on our previous derivation we update $Q(S_t, A_t)$ as:

$$Q(S_t, A_t) \leftarrow Q(S_t, A_t) + \alpha(R_{t+1} + \gamma Q(S_{t+1}, A_{t+1}) - Q(S_t, A_t)) \tag{9.13}$$

This expresses that the new value for an state-action pair is the old value plus how much the reward combined with the next state action value differs from the current Q-value. We add $\alpha$ times this difference.

Once we have these values for a policy $\pi$ we still need to decide on what action $A \in \mathcal{A}(S_t)$ to select. Various *action selection approaches* are available, including one called $\epsilon$-greedy. In the $\epsilon$-greedy approach we select a random action with probability $\epsilon$ to allow for exploration, while in all other cases we select the action with the highest value. This is formalized in Algorithm 25.

---

**Algorithm 25:** $\epsilon$-greedy action selection given a state $S$

---

r = random number from [0, 1]
**if** $r < \epsilon$ **then**
  |   **return** *a random action A from* $\mathcal{A}(S)$
**else**
  |   **return** $\mathrm{argmax}_{A \in \mathcal{A}(S)}\, Q(S, A)$

---

An alternative action selection mechanism is the *softmax* approach. The probability of selecting an actions $A$ in situation $S$ is defined as:

$$p(A|S) = \frac{e^{\frac{Q(A,S)}{\tau}}}{\sum_{A' \in \mathcal{A}(S)} e^{\frac{Q(A',S)}{\tau}}} \tag{9.14}$$

Here, $\tau$ expresses a *temperature*, similar to simulated annealing that we have treated before. $\tau$ has a value $> 0$. The lower the temperature, the more we select based on the value for $Q(A, S)$. The higher, the more we select in a random way. We can start with a high temperature (doing more exploration) and lower the temperature as we gather more experiences (exploitation).

---

**Algorithm 26:** Evolving a policy $\pi$ with SARSA

---

$\forall S \in \mathcal{S}, A \in \mathcal{A} : Q(S, A) = $ random
$S = \mathrm{derive\_state}(x_1)$
time = 1
Select an action $A$ based from $\mathcal{A}(S)$ based on our set of $Q(S, A)$'s using a selection approach
**while** *True* **do**
  |   Perform action $A$
  |   time = time + 1
  |   $S' = \mathrm{derive\_state}(x_{time})$
  |   R = observe reward
  |   Select an action $A'$ from $\mathcal{A}'(S')$ based on our set of $Q(S', A')$'s using a selection approach
  |   Perform action $A'$
  |   $Q(S, A) = Q(S, A) + \alpha(R + \gamma Q(S', A') - Q(S, A))$
  |   A = A'
  |   S = S'
**end**

---

**Table 9.1** Q(S, A) values for the Bruce case

|                            | Cheer up! (cu) | Remember: plan fun activity (pfa) |
| -------------------------- | -------------- | --------------------------------- |
| Signs of depression (sod)  | 20             | 65                                |
| Overly happy (oh)          | 0              | 0                                 |
| Reflecting (ref)           | 50             | 0                                 |

Given the selection method, we will explain the State-Action-Reward-State-Action algorithm SARSA to adjust the policy (note that we leave out some details with respect to the Markov model for the sake of simplicity). The algorithm is shown in Algorithm 26. We randomly assign a value for $Q(S, A)$ in the initial phase. We start by deriving the state $S$ from our initial set of sensory values $x_1$ and select an action based on some action selection mechanism and our values for all possible actions $Q(S, A)$. We perform the action and collect sensory data again to form state $S'$. We also observe the reward $R$. In this new situation, we select a next action $A'$. We can now update our value of $Q(S, A)$. We set the action $A$ to our new action $A'$, the state $S$ to $S'$ and continue in our loop by executing the action, observing the new state $S'$, etcetera.

Let us return to our previous example of Bruce. We eventually end up with the Q-values shown in Table 9.1. Note that we only show states where we can perform actions, thereby omitting the terminal states. Let us consider one example Q-value, namely $Q(sod, pfa)$ (considering the abbreviations noted in the table). Given our equation for the Q-values (assuming $\epsilon - greedy$) we compute that:

$$Q(sod, pfa) = Q(sod, pfa) + \alpha(15 + \gamma Q(ref, cu) - Q(sod, pfa)) \quad (9.15)$$

which results in a value of 65 in the end if a value of $\gamma = 1$ is selected.

We now have a simple approach to decide on feedback and support for our quantified selves. However, more sophistical approaches exist that help to improve this simple algorithm. We will discuss these next.

## 9.3   Q-Learning

Q-learning is certainly one of the most well-known reinforcement learning approaches and was introduced in [125]. In Q-learning we follow roughly the same approaches as we have seen in SARSA. However, while in SARSA we updated our values for $Q(S, A)$ based on the value of the selected action in the next state $S'$ using the same policy (e.g. $\epsilon$-greedy), for Q-learning we directly select the action that has the highest value for the next state. This simplifies our algorithm, as we can see below. The approach is sometimes called *off-policy* to reflect that we apply a different policy for the calculation of the new value of $Q(S, A)$ compared to the one we use to select the actual action $A$. SARSA is an *on-policy* approach as we use the same for both cases.

---

**Algorithm 27:** Evolving a policy with Q-learning

---
$\forall S \in \mathcal{S}, A \in \mathcal{A} : Q(S, A) = \text{random}$
$S = \text{derive\_state}(x_1)$
$\text{time} = 1$
**while** *True* **do**
    Select an action $A$ from $\mathcal{A}(S)$ based on our $Q(S, A)'s$ using a selection approach
    Perform action $A$
    $\text{time} = \text{time} + 1$
    $S' = \text{derive\_state}(x_{time})$
    $Q(S, A) = Q(S, A) + \alpha(R + \gamma \max_{A'(S')} Q(S', A') - Q(S, A))$
    $S = S'$
**end**

---

## 9.4  SARSA($\lambda$) and Q($\lambda$)

We are going to extend the approaches we have just considered. In our simple *SARSA* and *Q-learning* approaches we lacked a concept that was able to identify what actions we had taken in the past that contributed to the current state. This would allow us to put blame or credit to those choices we made before which we currently can only do for one step . These are so-called *eligibility traces*. Assume $Z_t(s, a)$ represents the eligibility trace at time point $t$ for state $s$ with action $a$. The value for the trace is defined as:

$$Z_t(s, a) = \begin{cases} \gamma \lambda Z_{t-1} + 1 & \text{if } s = S_t \wedge a = A_t \\ \gamma \lambda Z_{t-1} & \text{otherwise} \end{cases} \tag{9.16}$$

The combined factors $\gamma$ (as seen before) and $\lambda$ determine how quickly the history of the eligibility trace decays. We add 1 to the eligibility trace of a state if the state occurred and 0 otherwise. How do we incorporate this? Well, we update our equation for calculating $Q(S, A)$. For SARSA this becomes:

$$Q(S_t, A_t) \leftarrow Q(S_t, A_t) + \alpha \cdot (R_{t+1} + \gamma Q(S_{t+1}, A_{t+1}) - Q(S_t, A_t)) \cdot Z_t(S_t, A_t) \tag{9.17}$$

Here, we see that changes in the Q-value depend on the value of the eligibility trace. If the state-action pair is more eligible (i.e. in our history the pair has been applied more frequently), the magnitude of the update is increased. We can simply plug this into our learning algorithm we have defined before. For Q-learning the new equation becomes:

$$Q(S_t, A_t) = Q(S_t, A_t) + \alpha \cdot (R_{t+1} + \gamma \max_{A'(S_{t+1})} Q(S_{t+1}, A_{t+1}) - Q(S_t, A_t)) \cdot Z_t(S_t, A_t)$$

$$(9.18)$$

## 9.5   Approximate Solutions

So far we have assumed a value $Q(S, A)$ for each combination, but this is not very realistic in our quantified self setting. We are likely to have a lot of different states and hopefully also have quite some opportunities to provide a form of support or feedback (in other words: actions). We cannot just store these. We could however learn a *model* that predicts the value $Q(S, A)$ in a state $S$ for a given action $A$. Luckily, we have already seen a lot of these models before in our previous chapters. We can see each observation of a state $S$ and an action $A$ with a value $Q(S, A)$ as an instance of a learning problem with $Q(S, A)$ as our target. Assume we have a matrix of weights $\mathbf{w}$ that represents a model for our predictions. For instance, the weights of a neural networks. Assuming our model is represented as $\hat{f}(S_t, A_t, \mathbf{w})$ giving a prediction for $Q(S_t, A_t)$ we can define the error as:

$$\sum_{S \in S, A \in A} \sqrt{(Q(S_t, A_t) - \hat{f}(S, A, \mathbf{w}))^2} \qquad (9.19)$$

In case we do not have instances for all cases we obviously only consider the cases we have information about. We can apply this model to estimate the value for $Q(S, A)$ in our algorithms, and update the model after each state change.

## 9.6   Discretizing the State Space

One final aspect we will devote some attention to a way to discretize the state space. We could have an infinite state space in our quantified self setting, and we would like to create useful representations of states. If we just consider all possible values for the activity level of Arnold, we would already be lost with our previous approaches. For this purpose, the *U-tree algorithm* (cf. [119]) can be used. Note that we slightly simplify and generalize the approach to match with our previous algorithms. This approach dynamically discretizes the state space by means of a *state tree*, very much like the decision trees we have seen. The leaves of the tree represent a discrete state while the nodes tell us how to get there based on the continuous values observed in our current state.

We build the tree by starting with a single node (i.e. a leaf): all continuous states are mapped to one discrete state. We start a data collection phase (i.e. running a reinforcement learning algorithm) and store data about the states $S$ we have visited, the action $A$ selected, the next state $S'$ and the reward $R$. After having collected data for a while, we stop and consider the newly obtained values for $Q(S, A)$ for each of the leaves that we have (as said, initially we have one). For each leaf, we investigate whether a split could be beneficial. Hereto, we iterate over all attributes $X_i$ and sort the data according to the values of that attribute. We consider a split at each point in the ordered list of values and test for a significant difference between the Q-values in the two resulting sets using a Kolmogorov Smirnov test. We store both the p-value and the split criterion of the split with the lowest p-value (i.e. the most different subsets). In the end, we select the attribute with the lowest p-value. If the value is below 0.05 we split based on that attribute and the split value we have identified. That is all there is to it.

## 9.7 Exercises

### 9.7.1 Pen and Paper

1. It is known that reinforcement learning can be a slow learning algorithm. Imagine we have $u$ possible discrete input states and $v$ possible actions. In the worst case, how many inappropriate advices would be given to a user before the right one is provided, assuming there is only one right action for each input state?
2. Assume that we use data similar to our crowdsignals data. The objective is to make the user more active. What could be an appropriate definition of a reward function for a reinforcement learning algorithm we could deploy based on the data we have?
3. Let us continue on the case we briefly explained in the previous question: define a set of at least 5 actions that could be appropriate in terms of an intervention for this case.
4. A lot of the reinforcement learning algorithms revolve around the Markov property. Give an example of a setting in the quantified self where the Markov property is clearly not satisfied. Explain why the property is not satisfied.
5. We have seen both on-policy and off-policy approaches in this chapter. List an advantage of the on-policy approach towards reinforcement learning and list one disadvantage (as opposed to an off-policy approach).
6. Is the state space discretization approach that has been discussed in this chapter always suitable, or can you think of scenarios where it might be completely off? Argue why (not).

## 9.7.2   Coding

1. Implement one of the reinforcement learning algorithms we have seen in this chapter. The implementation should be able to handle continuous inputs, but only discrete outputs.
2. Select one of the datasets that are available (or one you can measure yourself). Define a setting of a user with a certain goal related to this dataset, and think of possible interventions you could perform (in this case: messages you can send) to support the user in achieving the goal. Implement a reward function that expresses in how far the user achieved the goal based on the data that has been collected.
3. Implement this system you developed under (2) on the mobile phone and evaluate how well this works, write a report on the experiences both from a machine learning perspective and the perspective of the user (did the user see improvement over time, were the suggestions useful, etcetera).

# Part III
# Discussion

# Chapter 10
# Discussion

Sadly enough we have reached the final chapter of this book. In this chapter, we aim to provide an outlook towards the future and discuss the challenges that we see for this domain. We have covered a lot of different topics within this book, where some of the topics we covered are not yet common practice in the domain of machine learning for the quantified self. Examples are the reinforcement learning techniques, the temporal predictive modeling techniques, and the outlier detection algorithms. Hence, even some parts we have described already still require a thorough evaluation. In addition, we will identify a number of issues that are not covered with the techniques we have explained in this book and that require additional research in terms of algorithmic developments. This is not meant to be an exhaustive list, but rather to give an idea on some developments we foresee will be required to advance the domain.

## 10.1 Learning Full Circle

Predictive modeling is a common research topic related to the quantified self, for instance the recognition of activities based on the sensory values. What is not common at all is learning how to use these insights to support the user in a personalized way, and the development of techniques to do so. As said, we suggest that the domain of reinforcement learning is a promising approach for this, but there are a number of issues that need to be addressed before these techniques can be used in practical setting:

1. *learning quickly*: the users will lose interest if interventions or feedback provided are not in line with the expectations and characteristics of the user. Arnold will not be happy in case he is provided with suggestions that would be suitable for his grandmother. The consequence is that the learning algorithm does not have endless opportunity to figure out what works for a user, and hence, it needs to

© Springer International Publishing AG 2018
M. Hoogendoorn and B. Funk, *Machine Learning for the Quantified Self*,
Cognitive Systems Monographs 35, https://doi.org/10.1007/978-3-319-66308-1_10

learn rapidly. Reinforcement learning is known to be slow. Therefore, we should create algorithms that learn faster, e.g. by exploiting multiple similar users at the same time. This opens up a whole range of interesting research questions: when are users similar? Should we consider their basic socio-demographic data? Or should we look at their responses to feedback and interventions? And how do we share the burden between users? Should we try out different interventions across the different users? One shot learning (see e.g. [45]) could be a promising approach as well.

2. *learning safely*: while learning fast is desirable, we might have to do with users that are vulnerable, such as Bruce. We do not want to provide Bruce with continuous suggestions that might cause him to become depressed again. Of course, we do want to figure out what works well for Bruce and what does not. These are two opposing forces: we do not want to constrain the search space for what intervention or feedback to provide, while we do want to constrain it to avoid doing harmful things. Exactly how to constrain algorithms to learn in this way is something that needs to be explored. There is some work on constrained reinforcement learning already [65] but more work is still required.

3. *using future predictions*: a lot of emphasis in this book has gone into predictive modeling. While this is certainly of value for non-intervention settings, things become blurry when we predict the future and intervene at the same time. How can we predict what would have happened if we did not intervene? And does the predictive modeling help us to intervene pro-actively and avoid undesired situations (e.g. Bruce having a nervous breakdown, Arnold loosing his shape)? These are issues that require rigorous evaluation, also with users.

## 10.2   Heterogeneity

Heterogeneity is a key phrase within the quantified self. We face heterogeneous users as well as heterogeneous devices, and even heterogeneity in the amount of devices a user might cary.

1. *learn across devices*: we should be able to perform machine learning over multiple devices with different specifications and capabilities. This would require mapping datasets to a more abstract level that is device independent, e.g. scale accelerometer values, use proxies for sensors that are not available on a certain device, etcetera. Precisely how to do this is an open research question. An example of a study that explored different phone platforms can be found in [21].

2. *learn across people*: we are potentially facing a lot of different users with their own characteristics. People have different walking speeds, carry their devices at different positions, have different preferences for support, etcetera. As argued before, learning fully individually is not always possible due to a lack of data. A challenge is therefore to learn generic models across people that are still reasonably accurate and can act as a starting point when there is insufficient data.

3. *coordinate behavior*: if a user carries multiple devices, then there should be a form of coordination between the devices. For instance, if we provide feedback, which device should we use at what time? Should we provide feedback on a smart watch when a user is in a conversation or a meeting? Of course, we do not want to show the same message on two devices at the same time. This context-dependent usage of different platforms and learning to coordinate between them is a direction that will require more and more attention.

## 10.3  Effective Data Collection and Reuse

Annotated data for the quantified self can be difficult to obtain. We should not require the user to insert lots of information without seeing an immediate benefit. To tackle this problem, not only can we learn across users or devices, but we can also improve the way in which data is collected.

1. *collection of data*: when learning is performed, some cases are much more interesting to label than others. If you observe data which clearly marks a particular activity that has been seen before there is no need to bother the user. On the other hand, if data that is completely different from what has been seen so far we might be very interested in knowing the label. The field of active learning (see [104] for a nice overview) could have great potential for this purpose.
2. *transfer between use cases*: for the quantified self setting we would expect to not just focus on learning one specific task (e.g. activity recognition) but to tackle multiple tasks. A question that arises is how we can reuse lessons learned across these tasks. Transfer learning (see e.g. [92]) would be a field that is useful in this respect.

## 10.4  Data Processing and Storage

We did not touch upon the data storage and efficient processing in this book at all, except when we discussed more efficient streaming data mining algorithms that avoid having to store and process entire datasets. Of course, it does pose interesting trade-offs and challenges that still require further investigation:

1. *storing data*: storing data in case we do not use a streaming approach needs to be done somewhere. Questions that arise are: how do we efficiently store the data? Where do we do this? Should this be locally on the phone, or somewhere in a cloud based infrastructure? There are systems that focus on this issue and try to manage this problem, see e.g. [69]. Of course, the choice for storage is highly intertwined with the algorithms being used: does it need to learn for individual users or across people?

2. *processing data*: once we have data available, there are trade-offs about where
   the processing should take place: where do we learn? Should this be done locally
   on the phone or in the cloud? And how often should we update our models? If we
   do not do this frequently enough we might not have an accurate representation
   of the user.
3. *battery management*: measuring everything we can as often as possible not only
   poses challenges for storage, but also for the battery of the phone: we potentially
   drain it quickly if we measure too often. In addition, the more data, the higher the
   complexity of the learning process, but also the more accurate it could potentially
   become. We should therefore develop algorithms that take the battery usage into
   account.

## 10.5   Better Predictive Modeling and Clustering

While we have explained a variety of approaches in the machine learning domain
that can contribute to predicting unknown values about a user, there is still room
for improvement in the context of the quantified self. Below, we list a number of
directions we feel would be promising.

1. *better features with less effort*: the identification of features can already be auto-
   mated to a large extent but is still considered more an art than following a scientific
   recipe. Sensors have different sampling rates and somehow we need to exploit all
   data in the best possible way. Deep learning is known for its ability to automatically extract useful features, we feel this is a promising avenue that should be
   explored further. In addition, extraction of temporal features for the non-temporal
   learning algorithms is a direction that also deserves more attention.
2. *domain knowledge*: while this book is all about machine learning, we should also
   consider the fact that domain knowledge can be extremely useful. We should
   not reinvent the wheel. Combining machine learning approaches with domain
   knowledge is in our opinion very important in the context of the quantified self,
   it can also help us to handle the cold start problem. We have already shown this
   a bit when we discussed the dynamical systems models in Chap. 8.
3. *temporal learning*: we have explained a number of temporal learning algorithms
   in this book, but many more exist. We feel that the temporal developments should
   result in better predictions than the ones we displayed in our case study. Recently,
   there have been developments in the area of temporal learning that have not
   been described in this book, e.g. LSTM [60], GRU [37], etc. We foresee more
   developments in this area, also for learning well across different users.

4. *explainability of models*: black box methods can often work pretty well in terms of predictive performance. However, a level of explainability of the model can also be desirable, for example if understanding the basis for an advice is of vital importance (e.g. when a therapist is involved to battle Bruce's notorious depressions). Trying to develop methods that explain exactly what features black box models use could be beneficial. See e.g. [132]

## 10.6  Validation

Although we have focused a lot on evaluation of our predictive modeling techniques, we did not focus on the validation of full-fletched systems that incorporate the techniques we have explained in this book. There are a number of issues that require attention to perform such a validation study that have in our opinion not been thoroughly addressed yet:

1. *do validation*: a lot of applications are seen (e.g. in the app stores) that make all sorts of claims without showing any form of proof that the app actually works (see e.g. [84]). We feel that especially health related apps should be much more rigorously evaluated before exposing users to them.
2. *definition of success*: the outcome measure is obviously highly dependent on the specific domain in which the app has been developed. If we for example develop an app for a specific disease (e.g. depression) or health goal, well-known measures are present that define success of a treatment, in our case being the app. However, these goals might not always be as clear. Possibly user engagement is more important, especially for companies selling apps. How to precisely define such metrics is still a challenge.
3. *setup of validation study*: for medical or health treatments very clear setups for validation studies exist, such as randomized controlled trials. These are rigorous studies with well defined protocols that take a long time to prepare and get approved. Based on our experience, these more traditional studies slow down the validation process to such an extent that the application under evaluation is already outdated when the actual trial starts. There is really a need for new paradigms that are faster, but still take the considerations of the users and privacy issues throughly into account. A/B testing is used frequently for evaluating websites and user behavior when browsing. Possibly a nice middle ground can be found.

# References

1. Aggarwal, C.C. (ed.): Data streams: models and algorithms. Springer Science & Business Media, New York (2007)
2. Abu-Mostafa, Y.S., Magdon-Ismail, M., Lin, H.T.: Learning from Data, vol. 4. AMLBook, Singapore (2012)
3. Aerts, M., Claeskens, G., Hens, N., Molenberghs, G.: Local multiple imputation. Biometrika, 375–388 (2002)
4. Agrawal, R., Gehrke, J., Gunopulos, D., Raghavan, P.: Automatic subspace clustering of high dimensional data. Data Min. Knowl. Discov. **11**(1), 5–33 (2005). doi:10.1007/s10618-005-1396-1
5. Agrawal, R., Srikant, R., et al.: Fast algorithms for mining association rules. In: Proceedings of 20th International Conference Very Large Data Bases, VLDB, vol. 1215, pp. 487–499 (1994)
6. Allen, J.F.: Maintaining knowledge about temporal intervals. Commun. ACM **26**(11), 832–843 (1983)
7. Anguita, D., Ghio, A., Oneto, L., Parra, X., Reyes-Ortiz, J.L.: A public domain dataset for human activity recognition using smartphones. In: European Symposium on Artificial Neural Networks, Computational Intelligence and Machine Learning (April), pp. 24–26 (2013). http://www.i6doc.com/en/livre/?GCOI=28001100131010
8. Anguita, D., Ghio, A., Oneto, L., Parra, X., Reyes-Ortiz, J.L.: Human activity recognition on smartphones using a multiclass hardware-friendly support vector machine. Lecture Notes in Computer Science (including subseries Lecture Notes in Artificial Intelligence and Lecture Notes in Bioinformatics). LNCS, vol. 7657, pp. 216–223 (2012)
9. Augemberg, K.: Building that perfect quantified self app: notes to developers, part 1. The Measured Me Blog (2012)
10. Banos, O., ttila Toth, M.A., Damas, M., Pomares, H., Rojas, I.: Dealing with the effects of sensor displacement in wearable activity recognition. Sensors (Basel, Switzerland) **14**(6), 9995–10,023 (2014). doi:10.3390/s140609995
11. Bao, L., Intille, S.S.: Activity Recognition from user-annotated acceleration data. Pervasive Comput., 1–17 (2004). doi:10.1007/b96922, http://www.springerlink.com/content/9aqflyk4f47khyjd
12. Batal, I., Valizadegan, H., Cooper, G.F., Hauskrecht, M.: A temporal pattern mining approach for classifying electronic health record data. ACM Trans. Intell. Syst. Technol. (TIST) **4**(4), 63 (2013)
13. Berchtold, M., Budde, M., Schmidtke, H.R., Beigl, M.: An extensible modular recognition concept that makes activity recognition practical. In: Annual Conference on Artificial Intelligence, pp. 400–409. Springer, Berlin (2010)
14. Berndt, D.J., Clifford, J.: Using dynamic time warping to find patterns in time series. KDD Workshop, Seattle, WA **10**, 359–370 (1994)

M. Hoogendoorn and B. Funk, *Machine Learning for the Quantified Self*,
Cognitive Systems Monographs 35, https://doi.org/10.1007/978-3-319-66308-1

15. Bhattacharya, S., Lane, N.D.: From smart to deep : robust activity recognition on smart-watches using deep learning. In: The Second IEEE International Workshop on Sensing Systems and Applications Using Wrist Worn Smart Devices (2016). doi:10.1109/PERCOMW. 2016.7457169
16. Bhattacharya, S., Nurmi, P., Hammerla, N., Plötz, T.: Using unlabeled data in a sparse-coding framework for human activity recognition. Pervasive Mob. Comput. **15**, 242–262 (2014). doi:10.1016/j.pmcj.2014.05.006
17. Biau, G., Devroye, L.: Lectures on the Nearest Neighbor Method. Springer, Berlin (2015)
18. Bishop, C.M.: Pattern Recognition and Machine Learning. Springer, Berlin (2006)
19. Blanke, U., Schiele, B.: Sensing location in the pocket. In: Ubicomp Poster Session, pp. 4–5 (2008). http://www.ulfblanke.de/research/ubicomp08/ubicomp08_paper_web.pdf
20. Blei, D.M., Ng, A.Y., Jordan, M.I.: Latent dirichlet allocation. J. Mach. Learn. Res. **3**, 993–1022 (2003)
21. Blunck, H., Jensen, M.M., Sonne, T., Science, C.: ACTIVITY RECOGNITION ON SMART DEVICES: Dealing with diversity in the wild. GetMobile **20**(1), 34–38 (2016)
22. Bogue, R.: Recent developments in mems sensors: a review of applications, markets and technologies. Sens. Rev. **33**(4), 300–304 (2013)
23. Bosse, T., Hoogendoorn, M., Klein, M.C., Treur, J.: An ambient agent model for monitoring and analysing dynamics of complex human behaviour. J. Ambient Intell. Smart Environ. **3**(4), 283–303 (2011)
24. Bosse, T., Hoogendoorn, M., Klein, M.C., Treur, J., Van Der Wal, C.N., Van Wissen, A.: Modelling collective decision making in groups and crowds: integrating social contagion and interacting emotions, beliefs and intentions. Auton. Agents Multi-Agent Syst. **27**(1), 52–84 (2013)
25. Both, F., Hoogendoorn, M., Klein, M.C., Treur, J.: Modeling the dynamics of mood and depression. In: ECAI, pp. 266–270 (2008)
26. Box, G.E., Jenkins, G.M., Reinsel, G.C., Ljung, G.M.: Time Series Analysis: Forecasting and Control, 5th edn. Wiley, Hoboken (2015)
27. Bracewell, R.: The fourier transform and its applications (1965)
28. Breda, W.v., Hoogendoorn, M., Eiben, A., Andersson, G., Riper, H., Ruwaard, J., Vernmark, K.: A feature representation learning method for temporal datasets. In: IEEE SSCI 2016. IEEE (2016)
29. Breiman, L.: Bagging predictors. Mach. Learn. **24**(2), 123–140 (1996)
30. Breunig, M.M., Kriegel, H.P., Ng, R.T., Sander, J.: Lof: identifying density-based local outliers. In: ACM Sigmod Record, vol. 29, pp. 93–104. ACM (2000)
31. Brockwell, P., Davis, R.: Introduction to Time Series and Forecastin. Springer, Berlin (2010)
32. Casale, P., Pujol, O., Radeva, P.: Human activity recognition from accelerometer data using a wearable device. In: Pattern Recognition and Image Analysis, pp. 289–296 (2011). doi:10. 1007/978-3-642-21257-4, doi:10.1007/978-3-642-21257-4_36
33. Chapelle, O., Schölkopf, B., Zien, A.: Semi-Supervised Learning. The MIT Press, Cambridge (2010)
34. Chatfield, C.: The Analysis of Time Series-An Introduction. Chapman & Hall, London (2004)
35. Chauvenet, W.: A Manual of Spherical and Practical Astronomy, vol. 1, 5th ed., revised and corr. Dover Publication, New York (1960)
36. Chen, Z., Lin, M., Chen, F., al, E.: Unobtrusive sleep monitoring using smartphones. In: Proceedings of the 11th ACM Conference on Embedded Networked Sensor Systems, pp. 4:1–4:14 (2013). doi:10.1145/2517351.2517359
37. Cho, K., Van Merriënboer, B., Bahdanau, D., Bengio, Y.: On the properties of neural machine translation: encoder-decoder approaches. arXiv preprint arXiv:1409.1259 (2014)
38. Choe, E.K., Lee, N.B., Lee, B., Pratt, W., Kientz, J.A.: Understanding quantified-selfers' practices in collecting and exploring personal data. In: Proceedings of the 32nd Annual ACM Conference on Human Factors in Computing Systems, pp. 1143–1152 (2014). doi:10.1145/ 2556288.2557372, http://dl.acm.org/citation.cfm?id=2557372
39. Cortes, C., Vapnik, V.: Support-vector networks. Mach. Learn. **20**(3), 273–297 (1995)

40. Deb, K., Agrawal, S., Pratap, A., Meyarivan, T.: A fast elitist non-dominated sorting genetic algorithm for multi-objective optimization: Nsga-ii. In: International Conference on Parallel Problem Solving From Nature, pp. 849–858. Springer, Berlin (2000)
41. Domingos, P., Hulten, G.: Mining high-speed data streams. In: Proceedings of the Sixth ACM SIGKDD International Conference on Knowledge Discovery and Data Mining, pp. 71–80. ACM (2000)
42. Duda, R.O., Hart, P.E., Stork, D.G.: Pattern Classification. Wiley, Cambridge (2012)
43. Durbin, J., Koopman, S.J.: Time Series Analysis by State Space Methods, vol. 38. OUP, Oxford (2012)
44. Eiben, A.E., Smith, J.E.: Introduction to Evolutionary Computing, 2nd edition, Springer (2015). doi:10.1007/978-3-662-44874-8
45. Fei-Fei, L., Fergus, R., Perona, P.: One-shot learning of object categories. IEEE Trans. Pattern Anal. Mach. Intell. **28**(4), 594–611 (2006)
46. Feldman, R., Sanger, J.: The Text Mining Handbook: Advanced Approaches in Analyzing Unstructured Data. Cambridge University Press, Cambridge (2007)
47. Fox, S., Duggan, M.: Tracking for health (2013). http://www.pewinternet.org/2013/01/28/main-report-8/
48. Fraden, J.: Handbook of Modern Sensors, vol. 3. Springer, Berlin (2010)
49. Freund, Y., Schapire, R.E.: A desicion-theoretic generalization of on-line learning and an application to boosting. In: European Conference on Computational Learning Theory, pp. 23–37. Springer, Berlin (1995)
50. GfK: A third of people track their health or fitness (2016). http://www.gfk.com/insights/press-release/a-third-of-people-track-their-health-or-fitness-who-are-they-and-why-are-they-doing-it/
51. Gimpel, H., Nißen, M., Görlitz, R.A.: Quantifying the quantified self: a study on the motivation of patients to track their own health. ICIS **2013**, 128–133 (2013)
52. Grubbs, F.E.: Sample criteria for testing outlying observations. Ann. Math. Stat., 27–58 (1950)
53. Grubbs, F.E.: Procedures for detecting outlying observations in samples. Technometrics **11**(1), 1–21 (1969)
54. Gu, F., Kealy, A., Khoshelham, K., Shang, J.: User-independent motion state recognition using smartphone sensors. Sensors **15**(12), 30636–30652 (2015)
55. Guha, S., Mishra, N., Motwani, R., O'Callaghan, L.: Clustering data streams. In: proceedings of the 41st Annual Symposium on Foundations of Computer Science, 2000, pp. 359–366. IEEE (2000)
56. Hao, T., Xing, G., Zhou, G.: isleep: Unobtrusive sleep quality monitoring using smartphones. In: Proceedings of the 11th ACM Conference on Embedded Networked Sensor Systems, SenSys '13. ACM (2013). doi:10.1145/2517351.2517359
57. Hastie, T., Tibshirani, R., Friedman, J.: The Elements of Statistical Learning (2013). doi:10.1007/b94608
58. Haykin, S.S.: Neural Networks and Learning Machines, vol. 3. Pearson Upper Saddle River, NJ, USA (2009)
59. He, Z., Xu, X., Deng, S.: Discovering cluster-based local outliers. Pattern Recognit. Lett. **24**(9), 1641–1650 (2003)
60. Hochreiter, S., Schmidhuber, J.: Long short-term memory. Neural comput. **9**(8), 1735–1780 (1997)
61. Hodge, V.J., Austin, J.: A survey of outlier detection methodologies. Artif. Intell. Rev. **22**(2), 85–126 (2004)
62. Jaeger, H.: Tutorial on training recurrent neural networks, covering BPPT, RTRL, EKF and the "echo state network" approach. GMD-Forschungszentrum Informationstechnik (2002)
63. Jaeger, H., Haas, H.: Harnessing nonlinearity: predicting chaotic systems and saving energy in wireless communication. Science **304**(5667), 78–80 (2004)
64. Jolliffe, I.: Principal Component Analysis. Wiley Online Library, Cambridge (2002)
65. Junges, S., Jansen, N., Dehnert, C., Topcu, U., Katoen, J.P.: Safety-constrained reinforcement learning for mdps. In: International Conference on Tools and Algorithms for the Construction and Analysis of Systems, pp. 130–146. Springer, Berlin (2016)

66. Kalman, R.E.: A new approach to linear filtering and prediction problems. J. Basic Eng. **82**(1), 35–45 (1960)
67. Kaufman, L., Rousseeuw, P.: Clustering by means of medoids. In: Dodge, Y. (ed.) Statistical Data Analysis Based on the L1 Norm, pp. 405–416. Springer, Berlin (1987)
68. Kaufman, L., Rousseeuw, P.J.: Finding Groups in Data: An Introduction to Cluster Analysis, vol. 344. Wiley, Cambridge (2009)
69. Kemp, R., Palmer, N., Kielmann, T., Bal, H.: The smartphone and the cloud: power to the user. International Conference on Mobile Computing. Applications, and Services, pp. 342–348. Springer, Berlin (2010)
70. Keogh, E., Ratanamahatana, C.A.: Exact indexing of dynamic time warping. Knowl. Inf. Syst. **7**(3), 358–386 (2005)
71. Kirkpatrick, S., Gelatt, C.D., Vecchi, M.P., et al.: Optimization by simmulated annealing. science **220**(4598), 671–680 (1983)
72. Knorr, E.M., Ng, R.T.: Algorithms for mining distancebased outliers in large datasets. In: Proceedings of the International Conference on Very Large Data Bases, pp. 392–403 (1998)
73. Kolmogorov, A.N.: Sulla determinazione empirica di una legge di distribuzione. na (1933)
74. Könönen, V., Mantyärrvi, J., Similä, H., Pärkkä, J., Ermes, M.: Automatic feature selection for context recognition in mobile devices. Pervasive Mob. Comput. **6**(2), 181–197 (2010). doi:10.1016/j.pmcj.2009.07.001
75. Kop, R., Hoogendoorn, M., ten Teije, A., Büchner, F.L., Slottje, P., Moons, L.M., Numans, M.E.: Predictive modeling of colorectal cancer using a dedicated pre-processing pipeline on routine electronic medical records. Comput. Biol. Med. **76**, 30–38 (2016)
76. Lane, N.D., Miluzzo, E., Lu, H., Peebles, D., Choudhury, T., Campbell, A.T.: A survey of mobile phone sensing. IEEE Commun. Mag. **48**(9), 140–150 (2010). doi:10.1109/MCOM. 2010.5560598
77. Lang, T., Rettenmeier, M.: Understanding consumer behavior with recurrent neural networks. In: Proceedings of MLRec, vol. 2 (2017)
78. Lara, O.D.: Labrador, M.A.: A survey on human activity recognition using wearable sensors. IEEE Commun. Surv. Tutor. **15**(3), 1192–1209 (2013). doi:10.1109/SURV.2012.110112. 00192, http://ieeexplore.ieee.org/lpdocs/epic03/wrapper.htm?arnumber=6365160
79. LeCun, Y., Bengio, Y., Hinton, G.: Deep learning. Nature **521**(7553), 436–444 (2015)
80. Li, I., Dey, A., Forlizzi, J.: Understanding My Data, Myself: Supporting self-reflection with Ubicomp technologies. In: Proceedings of the 13th International Conference on Ubiquitous Computing, pp. 405–414 (2011). doi:10.1145/2030112.2030166, http://dl.acm.org/citation. cfm?id=2030166
81. Liao, T.W.: Clustering of time series dataa survey. Pattern Recognit. **38**(11), 1857–1874 (2005)
82. Lloyd, S.: Least squares quantization in pcm. IEEE Trans. Inf. Theory **28**(2), 129–137 (1982)
83. Lupton, D.: Self-tracking modes: reflexive self-monitoring and data practices. In: Imminent Citizenships: Personhood and Identity Politics in the Informatic Age, August (2014)
84. Middelweerd, A., Mollee, J.S., van der Wal, C.N., Brug, J., te Velde, S.J.: Apps to promote physical activity among adults: a review and content analysis. Int. J. Behav. Nutr. Phys. Act. **11**(1), 97 (2014)
85. Mitchell, T.M.: Machine Learning. McGraw-Hill Science, New York (1997)
86. Mitsa, T.: Temporal Data Mining. CRC Press, Hoboken (2010)
87. Mohri, M., Rostamizadeh, A., Talwalkar, A.: Foundations of Machine Learning. MIT press, Cambridge (2012)
88. Muaremi, A., Arnrich, B., Tröster, G.: Towards measuring stress with smartphones and wearable devices during workday and sleep. BioNanoScience **3**(2), 172–183 (2013). doi:10.1007/ s12668-013-0089-2
89. Neff, G., Nafus, D.: Self-Tracking. The MIT Press, Cambridge (2016)
90. Niennattrakul, V., Ratanamahatana, C.A.: On clustering multimedia time series data using k-means and dynamic time warping. In: 2007 International Conference on Multimedia and Ubiquitous Engineering (MUE'07), pp. 733–738. IEEE (2007)

91. Novikoff, A.B.: On convergence proofs on perceptrons. In: Symposium on the Mathematical Theory of Automata, pp. 615–622. Polytechnic Institute of Brooklyn (1962)
92. Pan, S.J., Yang, Q.: A survey on transfer learning. IEEE Trans. Knowl. Data Eng. **22**(10), 1345–1359 (2010)
93. Pärkkä, J., Ermes, M., Korpipää, P., Mäntyjärvi, J., Peltola, J., Korhonen, I.: Activity classification using realistic data from wearable sensors. IEEE Trans. Inf. Technol. Biomed. Publ. IEEE Eng. Med. Biol. Soc. **10**(1), 119–128 (2006). doi:10.1109/TITB.2005.856863
94. Peterek, T., Penhaker, M., Gajdoš, P., Dohnálek, P.: Comparison of classification algorithms for physical activity recognition. In: Innovations in Bio-inspired Computing and Applications, pp. 123–131. Springer, Berlin (2014)
95. Pierce, D.A.: A duality between autoregressive and moving average processes concerning their least squares parameter estimates. Ann. Math. Stat. **41**(2), 422–426 (1970)
96. Quinlan, J.R.: Induction of decision trees. Mach. Learn. **1**(1), 81–106 (1986)
97. Quinlan, J.R.: Improved use of continuous attributes in c4. 5. J. Artif. Intell. Res. **4**, 77–90 (1996)
98. Rabbi, M., Ali, S., Choudhury, T., Berke, E.: Passive and in-situ assessment of mental and physical well-being using mobile sensors. In: Proceedings of the 13th International Conference on Ubiquitous Computing, pp. 385–394. ACM (2011)
99. Rojas, R.: Neural Networks: A Systematic Introduction. Springer Science & Business Media, Berlin (2013)
100. Rosenblatt, F.: The perceptron: a probabilistic model for information storage and organization in the brain. Psychol. Rev. **65**(6), 386 (1958)
101. Rousseeuw, P.J.: Silhouettes: a graphical aid to the interpretation and validation of cluster analysis. J. Comput. Appl. Math. **20**, 53–65 (1987)
102. Rummery, G.A., Niranjan, M.: On-line Q-learning Using Connectionist Systems. University of Cambridge, Department of Engineering, Cambridge (1994)
103. Salton, G., Wong, A., Yang, C.S.: A vector space model for automatic indexing. Commun. ACM **18**(11), 613–620 (1975)
104. Settles, B.: Active Learning Literature Survey, vol. 52, Issue no. 11, pp. 55–66. University of Wisconsin, Madison (2010)
105. Shalev-Shwartz, S., Ben-David, S.: Understanding Machine Learning: From Theory to Algorithms. Cambridge University Press, Cambridge (2014)
106. Shannon, C.E.: Prediction and entropy of printed english. Bell Labs Tech. J. **30**(1), 50–64 (1951)
107. Shumway, R., Stoffer, D.: Time Series Analysis and Its Applications. Springer, Berlin (2011)
108. Smola, A., Vapnik, V.: Support vector regression machines. Adv. Neural Inf. Process. Syst. **9**, 155–161 (1997)
109. Song, M., Wang, H.: Highly efficient incremental estimation of gaussian mixture models for online data stream clustering. In: Proceedings of SPIE Conference, vol. 5803, p. 175 (2005)
110. Statista: Number of connected wearable devices worldwide (2016). https://www.statista.com/statistics/487291/global-connected-wearable-devices/
111. Sun, S.L., Deng, Z.L.: Multi-sensor optimal information fusion kalman filter. Automatica **40**(6), 1017–1023 (2004)
112. Sutton, R.S., Barto, A.G.: Reinforcement Learning: An Introduction. MIT press, Cambridge (1998)
113. Swan, M.: Sensor Mania! The internet of things, wearable computing, objective metrics, and the quantified self 2.0. J. Sens. Actuator Netw. **1**(3), 217–253 (2012). doi:10.3390/jsan1030217, http://www.mdpi.com/2224-2708/1/3/217/htm
114. Swan, M.: The quantified self: fundamental disruption in big data science and biological discovery. Big Data **1**(2), 85–99 (2013). doi:10.1089/big.2012.0002
115. Takagi, M., Fujimoto, K., Kawahara, Y., Asami, T.: Detecting hybrid and electric vehicles using a smartphone. In: Proceedings of 2014 ACM International Joint Conference on Pervasive and Ubiquitous Computing, pp. 267–275 (2014). doi:10.1145/2632048.2632088

116. Tapia, E.M., Intille, S.S., Haskell, W., Larson, K., Wright, J., King, A., Friedman, R.: Real-time recognition of physical activities and their intensities using wireless accelerometers and a heart monitor. In: Proceedings of the International Symposium on Wearable Comp (2007)

117. Tibshirani, R.: Regression shrinkage and selection via the lasso. J. Royal Stat. Soc. Ser. B (Methodological), 267–288 (1996)

118. Transition-Aware Human Activity Recognition using smartphones: Reyes-Ortiz, J.L., Oneto, L., Sama, A., Parra, X., https://orcid.org/0000-0002-4943-3021 Anguita, D.A.I.O. Neurocomput. Int. J. **171**, 754–767 (2016). doi:10.1016/j.neucom.2015.07.085, http://ovidsp.ovid.com/ovidweb.cgi?T=JS&PAGE=reference&D=psyc11&NEWS=N&AN=2015-39180-001

119. Uther, W.T., Veloso, M.M.: Tree based discretization for continuous state space reinforcement learning. In: Aaai/iaai, pp. 769–774 (1998)

120. van Breda, W.R., Hoogendoorn, M., Eiben, A., Berking, M.: An evaluation framework for the comparison of fine-grained predictive models in health care. In: Conference on Artificial Intelligence in Medicine in Europe, pp. 148–152. Springer, Berlin (2015)

121. Vapnik, V.N.: Statistical Learning Theory, vol. 1. Wiley, New York (1998)

122. Vapnik, V., Chervonenkis, A.: On the uniform convergence of relative frequencies of events to their probabilities. Theory Probab. Appl. **16**(2), 264–280 (1971)

123. Wang, H., Fan, W., Yu, P.S., Han, J.: Mining concept-drifting data streams using ensemble classifiers. In: Proceedings of the Ninth ACM SIGKDD International Conference on Knowledge Discovery and Data Mining, pp. 226–235. ACM (2003)

124. Ward Jr., J.H.: Hierarchical grouping to optimize an objective function. J. Am. Stat. Assoc. **58**(301), 236–244 (1963)

125. Watkins, C.J.C.H.: Learning from delayed rewards. Ph.D. thesis, University of Cambridge, England (1989)

126. Werbos, P.J.: Backpropagation through time: what it does and how to do it. Proc. IEEE **78**(10), 1550–1560 (1990)

127. Wiering, M., Van Otterlo, M.: Reinforcement learning. Adapt. Learn. Optim. **12** (2012)

128. Williamson, J., Liu, Q., Lu, F., Mohrman, W., Li, K., Dick, R., Shang, L.: Data sensing and analysis: challenges for wearables. In: 20th Asia and South Pacific Design Automation Conference, ASP-DAC 2015, pp. 136–141 (2015). doi:10.1109/ASPDAC.2015.7058994

129. Wu, W., Dasgupta, S., Ramirez, E.E., Peterson, C., Norman, G.J.: Classification accuracies of physical activities using smartphone motion sensors. J. Med. Internet Res. **14**(5), 1–9 (2012). doi:10.2196/jmir.2208

130. Zarchan, P.: Fundamentals of Kalman Filtering: A Practical Approach, 4th edn. AIAA (2015)

131. Zhang, M., Sawchuk, A.: USC-HAD: a daily activity dataset for Ubiquitous activity recognition using wearable sensors. In: Proceedings of ACM Ubiquitous Computing (UbiComp) (2012). doi:10.1145/2370216.2370438, http://www-scf.usc.edu/mizhang/papers/mi_ubicomp_sagaware12.pdf

132. Zintgraf, L.M., Cohen, T.S., Adel, T., Welling, M.: Visualizing deep neural network decisions: Prediction difference analysis. arXiv preprint arXiv:1702.04595 (2017)

133. Zou, H., Hastie, T.: Regularization and variable selection via the elastic net. J. Royal Stat. Soc. Ser. B (Statistical Methodology) **67**(2), 301–320 (2005)

# Index

© Springer International Publishing AG 2018
M. Hoogendoorn and B. Funk, *Machine Learning for the Quantified Self*,
Cognitive Systems Monographs 35, https://doi.org/10.1007/978-3-319-66308-1

Printed in the United States
By Bookmasters